ADVANCED LEVEL STATISTICS

ADVANCED LEVEL STATISTICS

Unit 3

CONDITIONAL PROBABILITY

Emrys Read
Y Coleg Normal, Bangor

PREFACE

This series of units is based on the current WJEC A Level statistics syllabus, but it may also be used to support other syllabi. The work has been organised into a total of ten units, each unit consisting of five modules, with each module corresponding to roughly a week's work. There is an additional computer unit and disk of programs which are linked to sections of the other units.

These units may be used either as additional material for teaching statistics in the context of a standard sixth form class or as the basis for a self-study course with tutor support. To serve this purpose, each module contains self-assessment questions in addition to tasks to be marked by the teacher or tutor. A separate work card is provided, which provides guidelines for the student concerning the content of the unit, suggestion for practical work, references to relevant parts of the computer unit and a list of additional resources (not provided as part of the pack) such as videos and other books.

All the resources are available in both Welsh and English.

The complete course consists of the following units:

- Unit 1: Descriptive Statistics
- Unit 2: Introduction to Probability
- Unit 3: Conditional Probability
- Unit 4: Discrete Distributions
- Unit 5: Continuous Distributions
- Unit 6: The Normal Distribution
- Unit 7: Joint Distributions
- Unit 8: Sampling Distributions and Point Estimation
- Unit 9: Interval Estimation
- Unit 10: Regression

Computer Unit

The units are published and distributed by:

Y Ganolfan Adnoddau Addysg
Faculty of Education
Yr Hen Goleg
Aberystwyth

Steering Committee Members

Chairman: James Nicholas HMI

Project Director: Gareth Roberts, Gwynedd Education Authority

Project Leader: Tom G. Edwards, Y Coleg Normal, Bangor

Members: George Barber, Clwyd Education Authority
I. Gwyn Evans, University College of Wales, Aberystwyth
John Wyn Jones, Ysgol Gyfun Llangefni
Emrys Read, Y Coleg Normal, Bangor
Harold Taylor, Yale Sixth Form College, Wrexham

© Crown Copyright 1989
Published with the permission of the Controller of Her Majesty's Stationery Office. These materials are the result of a research project funded by the Welsh Office Education Department. They have been prepared by Gwynedd Education Authority and are published by Y Ganolfan Adnoddau, University College of Wales, Aberystwyth.

UNIT 3

Contents

Module 1
1.1 Arrangements
1.2 Permutations
1.3 Combinations

Module 2
2.1 Introduction and development of the concept of conditional probability.
2.2 The multiplication formulae
2.3 The relationship with module 1
2.4 A useful result

Module 3
3.1 Intuitive definition of independence
3.2 Mathematical definition of independence
3.3 Problems involving 2 independent events
3.4 Independence of 3 events
3.5 Problems involving 3 or more independent events

Module 4
4.1 Probability Tree Diagrams
4.2 The Law of Total Probability
4.3 Bayes' Theorem

Module 5
5.1 Important results and formulae
5.2 Solutions to examination questions

Aims

1. To acquaint the reader with some of the methods which can be used for enumerating different kinds of arrangements and for calculating the probabilities which arise therefrom.
2. To introduce the reader to the concept of conditional probability and to develop methods for calculating such probabilities including the particular case of independent events.

Objectives

After completing the unit, you should be able to:

1. Calculate probabilities associated with various arrangements, including permutations and combinations.
2. Calculate simple conditional probabilities.
3. Calculate the probability of the occurence of 2, 3, or more events.
4. Decide when 2 or more events are independent and to calculate probabilities involved with such independent events.
5. Draw probability tree diagrams and use them to solve problems.
6. Calculate conditional probabilities by the use of Bayes' Theorem.

UNIT 3

Module 1

Arrangements, Permutations and Combinations

A knowledge of the size of a given sample space is usually essential when deriving probabilities, but writing down all the possible outcomes of an experiment can be a laborious and prohibitive task. In this module we shall develop methods which will in certain circumstances enable us to quickly calculate the total number of elements in a sample space, and thus to calculate the probabilities of events associated with that sample space.

1.1 Arrangements

We start this section by considering a single example

Example 1.1.1

Write down all the three figure numbers which can be formed from the digits {1,2,3}. Each digit may be used more than once in any number.

You should attempt this question yourself before reading any further.

In all you should have 27 different numbers, and these are best displayed by using a systematic approach. Firstly, we note that the first digit can be any one of three i.e. 1,2 or 3. Similarly, we have a choice of three for the second digit, and thus the first two digits may take any one of the nine forms

```
    1 1        2 1        3 1
    1 2        2 2        3 2
    1 3        2 3        3 3
```

Finally, we have a similar choice for the final digit, and the 27 possibilities are displayed below.

```
   1 1 1      2 1 1      3 1 1
   1 1 2      2 1 2      3 1 2
   1 1 3      2 1 3      3 1 3

   1 2 1      2 2 1      3 2 1
   1 2 2      2 2 2      3 2 2
   1 2 3      2 2 3      3 2 3

   1 3 1      2 3 1      3 3 1
   1 3 2      2 3 2      3 3 2
   1 3 3      2 3 3      3 3 3
```

An alternative method of recording these numbers is by using a tree diagram and is constructed as follows:

Stage 1

Write down all the values which the first figure can take

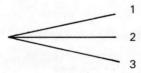

Stage 2

For each possible choice of first digit, write down all the possible values which the second digit can take.

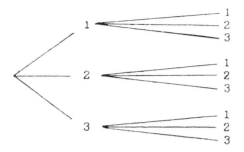

Stage 3

The diagram is now completed by filling in all the possibilities for the final digit

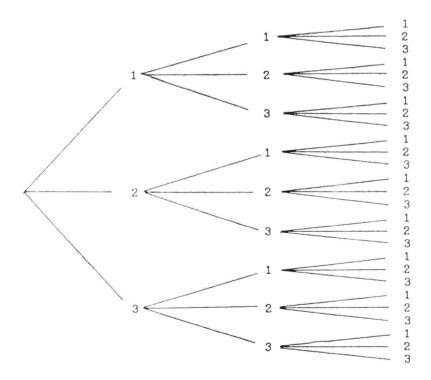

The individual 3 digit numbers can now be found by starting on the left hand side of the diagram and moving along any set of three branches to the right hand side. Indeed, once we have drawn the tree diagram, it is easy to appreciate why there must be in all 27 different possibilities. We have three choices at the first stage, and for **each** such choice, there are also three choices at the second stage. Thus, the first two digits may be chosen in 3x3 = 9 ways, and these have actually been noted above. The final digit may again be chosen in any one of three ways irrespective of what the first two digits may be, and thus there are 9x3 = 3x3x3 = 27 possible outcomes.

Example 1.1.2

Draw a tree diagram to calculate how many three figure numbers may be formed form the digits {1,2,3,4} if no digit may be used more than once in any one number.

Stage 1

Choose the first digit

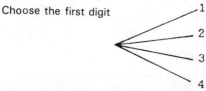

Stage 2

For each digit chosen at the first stage, write down all the possible values of the second digit (repeats are not allowed).

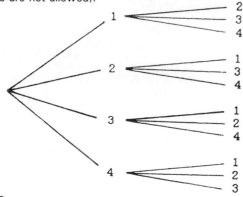

Stage 3

For each pair of digits which have already been chosen, write down all possible values of the final digit. Thus, if e.g. the first two digits are 2,4, the final digit can only be 1 or 3.

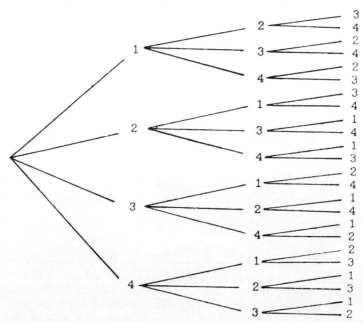

The total number of possible outcomes may now be calculated either by counting or more simply by noting that there are four choices at the first stage. For each such choice there are three possible choices at the second stage, and whatever the first two digits, there are always two choices at the final stage. Thus, altogether, there are 4x3x2 = 24 possible numbers.

Example 1.1.3

How many even numbers greater than 40,000 can be constructed from the digits {2,3,4,5,8} if no digit may be repeated more than once in any number.

Since the numbers are to be greater than 40,000, the first digit must be 4,5 or 8. Similarly for the number to be even, the final digit must be 2,4 or 8. Thus we shall consider the digits in the order 1st, 5th, 2nd, 3rd, 4th.

Stage 1

Choose the first digit

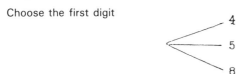

Stage 2

Choose the 5th digit

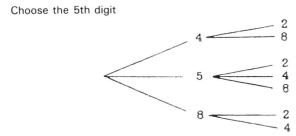

Stage 3

Choose the second digit. (We could now equally consider the 3rd or 4th digit)

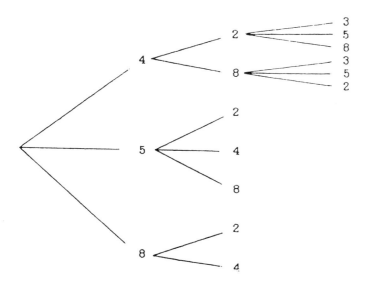

You should complete the 3rd stage for yourself.

Stage 4

Choose the 3rd digit

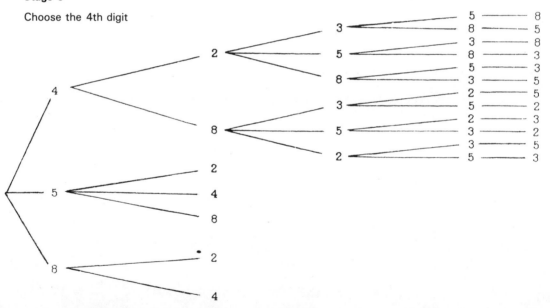

To be completed

Stage 5

Choose the 4th digit

You should now complete the diagram in its final form.

This time we have to be careful when enumerating the total number of possibilities. We note that if the first digit is a 4, then there are 2 choices at the second stage, 3 choices at the third stage, 2 at the fourth stage and 1 and the fifth stage, and thus in all 2x3x2x1 = 12 possibilities. Similarly there are 12 possibilities if the first digit is an 8. However, if the first digit is a 5, there are 3 choices at the second stage, 3 at the third stage, 2 at the fourth stage and 1 at the 5th stage, and hence altogether 3x3x2x1 = 18 possibilities. Thus the total number of possibilities is 12+12+18 = 42

Note

In drawing our tree diagrams, we first of all dealt with the first and fifth digit, since in both cases our choice was restricted by the information given in the question. The order in which we dealt with the remaining digits was unimportant. We could also have arrived at our result by first of all considering the fifth digit, then the first, and then the remaining three digits in any order.

Example 1.1.4

In how many ways can the letters of the word CAERGYBI be arranged?

In the previous example, we saw that drawing a tree diagram can be a laborious process. In this case, it would be prohibitive since it would consist of eight separate stages, but since we only need to calculate the total number of possible arrangements, we can now use the experience gained from drawing tree diagrams to argue as follows:

The first letter may be chosen in any one of 8 ways. For each choice of first letter, the second letter may be chosen in any one of 7 ways, and thus the first two letters may be chosen in any one of 8x7 ways. Similarly, for any given choice of the first two letters, the third letter may be chosen in any one of 6 ways, and therefore the first three letters may be chosen in 8x7x6 ways. Continuing this argument until we reach the final letter of which we only have one choice since it is the only letter remaining, we see that altogether we have

$$8 \times 7 \times 6 \times 5 \times 4 \times 3 \times 2 \times 1 = 8! \text{ possibilities}$$
$$= 40{,}320 \text{ possible arrangements}$$

[Note 8! is read as 'factorial 8']

Self Assessment Questions

You are recommended to draw tree diagrams to solve (at least) the first three questions.

1.1.5 In how many ways can any three of the letters {A,B,C,D} be arranged if
(i) a letter may be used more than once in any arrangement
(ii) no letter may be repeated in any arrangement

1.1.6 The diamond face cards (i.e. King, Queen, Jack) are removed from a pack of playing cards and then placed in a row from left to right. In how many different ways can this be done?

1.1.7 How many odd numbers greater than 70,000 can be formed from the digits {1,8,5,6,9} if
(i) repetitions are allowed
(ii) repetitions are not allowed

1.1.8 In how many ways can four of the letters {C,L,W,Y,D} be arranged if in any arrangement
(i) no letter may be repeated
(ii) repeats are allowed

1.1.9 A man who works a five day week can travel to work either by car or by bus. In how many different ways can he arrange a week's travelling to work?

1.1.10 How many arrangements are there of the letters of the word DINAS which start with a vowel?

1.1.11 I have 10 hard back books of identical size which I want to place on a shelf. There is however, only room for 6 of these books. In how many ways can I fill the 6 spaces if I take into account the order in which the books are placed.

Example 1.1.12

A 6th form chess club consists of 4 boys and 3 girls. One boy and one girl are to be chosen to represent the school in a competition, together with a reserve of each sex. In how many ways can this be done?

Solution 1.1.13 of example 1.1.12

Attempt this question by choosing suitable notation and writing down all the possible choices before reading the solution below:

There are in all 72 possible outcomes, and recording all of them is a laborious task. We can however arrive at the above number as follows. Let X, Y, Z denote the three girls and A, B, C, D the four boys. Then the girl chosen as first choice can be any one of three, and once the first choice has been made, the reserve girl can be any one of two. Thus the two girls can be chosen in 3x2 = 6 ways,
i.e.   XY   YX   ZX
        XZ   YZ   ZY

Here the first letter denotes the girl actually chosen and the second letter the reserve. Similarly the two boys may be chosen in 4×3 = 12 ways, i.e.

AB	BA	CA	DA
AC	BC	CB	DB
AD	BD	CD	DC

However, we can combine any choice of girls with any choice of boys, since the way in which the girls are chosen in no way affects the choice of boys thus

XYAB	XYBA	XYCA	XYDA
XYAC	XYBC	XYCB	XYDB
XYAD	XYBD	XYCD	XYDC

represents the 12 possible choices where the girls are XY. Similarly there will be 12 possible choices when the girls are XZ, and so on for each possible choice of girls. Thus altogether there must be 6×12 = 72 ways in which the two girls and two boys can be chosen.

Example 1.1.14

How many arrangements are there of the letters of the word PANAMA?

We note that in this example there are three As. For the moment, we shall denote them by A_1, A_2, and A_3, and taking these subscripts into account, we see that the letters may be arranged in 6×5×4×3×2×1 = 6! ways. However, it is clear that this is not the correct answer to our problem, since on removing the subscripts e.g.

$A_1 PMNA_2 A_3$ $A_2 PMNA_1 A_3$ $A_3 PMNA_1 A_2$
$A_1 PMNA_3 A_2$ $A_2 PMNA_3 A_1$ $A_3 PMNA_2 A_1$

will all give rise to APMNAA. Similarly MNPAAA may be derived from

$MNPA_1 A_2 A_3$ $MNPA_2 A_1 A_3$ $MNPA_3 A_1 A_2$
$MNPA_1 A_3 A_2$ $MNPA_2 A_3 A_1$ $MNPA_3 A_2 A_1$

If you were now to choose any unsubscripted arrangement and then write down the corresponding subscripted arrangements, it should be clear that in all cases, there are always six subscripted arrangements which give rise to the same unsubscripted arrangement. The reason for this is that within any such arrangement, the $\{A_1, A_2, A_3\}$ may be arranged amongst themselves without altering the unsubscripted form, and this can be done in 3×2×1 = 6 ways. Thus the original answer is 6 times too big and hence the correct answer is 6!/3! = 120 ways

Example 1.1.15

The letters of CAERGYBI are arranged at random. Find the probability that the first three letters will be C,R,G in some order.

We saw in the solution of 1.1.4 that there are altogether 8! possible arrangements. The number of arrangements which begin with the letters $\{C,R,G\}$ may be calculated as follows:

The first three letters of any such arrangement may be written down in 3×2×1 = 3! different ways. The final five letters must be $\{A,E,I,Y,B\}$ and these may be written down in 5×4×3×2×1 = 5! different ways. Since the choice of the order of the first three letters in no way affects the choice of the order of the final five, then in all there are 3!×5! arrangements which satisfy the required conditions. Thus the probability that an arrangement chosen at random will begin with the letters $\{C,R,G\}$ =

$$\frac{3! \times 5!}{8!} = \frac{3 \times 2 \times 1}{8 \times 7 \times 6} = \frac{1}{56}$$

Example 1.1.16

The letters of the word TREFOR are arranged at random. Find the probability that the two Rs are separated.
We shall exhibit two different methods for solving this problem.

Method 1

This involves calculating the total number of possible arrangements, and then subtracting the number of arrangements in which the two Rs appear next to one another. The total number of arrangements may be calculated as in example 1.1.14 and is seen to be $6!/2! = 360$. If we now think of these two Rs as being tied together giving us just $\{(RR), T,E,F,O\}$ to arrange, we see that there are $5 \times 4 \times 3 \times 2 \times 1 = 5! = 120$ arrangements in which the two Rs appear next to one another. Thus they will be separated in $360 - 120 = 240$ arrangements, and hence the probability that the two Rs are separated $= 240/360 = 2/3$.

Method 2

We first of all denote the two Rs by R_1 and R_2, and then place them on one side. The four remaining letters $\{F,T,O,E\}$ may be arranged amongst themselves in $4!$ ways $= 24$ ways, and let us assume that one such arrangement has been set out with spaces between the letters as shown below

$$\uparrow E \uparrow F \uparrow O \uparrow T \uparrow$$

R_1 and R_2 may now be inserted into any 2 of the 5 spaces denoted by the arrows in $5 \times 4 = 20$ ways, giving us an arrangement of the letters of the word TR_1EFOR_2 in which R_1 and R_2 are separated. As this can be done for any of the 24 arrangements of $\{F,T,O,E\}$ then altogether there must be $24 \times 20 = 480$ such arrangements. Since the number of possible arrangements of $\{T,R_1,E,F,O,R_2\}$ is equal to $6! = 720$ we see again that the required probability $= 480/720 = 2/3$.

Example 1.1.17

(i) In how many ways can Alun, Bryn, Carwyn, and Dafydd be seated around a round table?
(ii) Find the probability that in such an arrangement Alun and Dafydd are facing each other.

(i) We calculate the total number of possible arrangements in two ways.

Method 1

4 people can be arranged in $4! = 4 \times 3 \times 2 \times 1 = 24$ ways. However, since the table is round, any given position is repeated 4 times corresponding to rotations around the tables e.g.

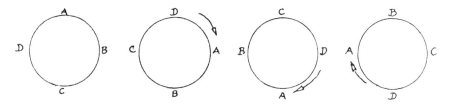

are all equivalent. Thus the number of different arrangements $= 4!/4 = 6$

Method 2

We are concerned with the positions of different people relative to one another, not relative to the table. Thus one person may be thought of as being fixed and then the other 3 being arranged around him, and hence there are $3! = 6$ different arrangements.
There are 2 of these 6 arrangements in which Alun and Dafydd face each other

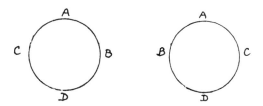

Thus the probability that Alun and Dafydd face each other $= 2/6 = 1/3$.

Self Assessment Questions

1.1.18 A head boy, head girl, deputy head boy and deputy head girl are to be chosen from a sixth form of 15 boys and 12 girls. In how many different ways can this be done?

1.1.19 A footall supporters' club decides to issue identity cards to its members. On each card there are to be 2 letters chosen from {A,B,C,D} followed by any 3 digits. What is the maximum number of members which the club can have if
(i) no digit or letter may be repeated on any identity card,
(ii) digits may be repeated but letters may not,
(iii) both letters and digits may be repeated.

1.1.20 How many arrangements are there of the letters of the words
(i) PENYBONT
(ii) ABERTAWE
(iii) BANANA

1.1.21 One of the arrangements of 1.1.20(ii) is chosen at random. Find the probability that
(i) the two As come together,
(ii) the R and T are separated.

1.1.22 One of the arrangements of 1.1.20 (iii) is chosen at random. Find the probability that all three As are separated.

1.1.23 (i) In how many different ways can 3 red counters 5 blue counters and 4 yellow counters be placed in a straight line.
(ii) Find the probability that in any such arrangement the first 3 counters will all be blue.

1.1.24 (i) In how many ways can 8 people be seated around a round table?
(ii) If only two of them are men, find the probability that in any such arrangement, the 2 men will be sitting next to one another.

1.1.25 (i) Calculate the number of dominoes in an ordinary set.
If one domino is removed at random from the set find the probability that
(ii) it will be a 'double',
(iii) the sum of the dots will be 5 or less.

1.1.26 (i) In how many ways can 8 beads, all of different colours, be arranged on a hoop (which may of course be turned over)?
(ii) If three of the beads are red, white and yellow, find the probability that in any such arrangement, these three beads will be adjacent to one another.

1.2 Permutations

Some of the questions which we have tackled essentially involve choosing a given number of objects form a given set and then arranging them in order. e.g.
How many four figure numbers can be formed from the digits {1,2,3,4,5,6} if no repeats are allowed?
We all know that the number of required arrangements is 6x5x4x3 and this can be rewritten in factorial notation

$$\frac{6 \times 5 \times 4 \times 3 \times 2 \times 1}{2 \times 1} = \frac{6!}{2!}$$

Arrangements of this kind are called permutations and further examples of permutations may be found in 1.1.2, 1.1.4, 1.1.5(ii), 1.1.6, 1.1.8(i), and 1.1.11. You should now return to these questions giving the answer in factorial notation in each case.
Can you suggest a general formula which gives the total number of ways of permuting r objects from a set of n unlike objects? Questions 1.1.2, 1.1.5(ii), 1.1.8, and 1.1.11 might suggest that n!/(n-r)! is the required result, but if we try and apply this result to e.g. 1.1.4 we get 8!/(8-8)! = 8!/0!, and 0! is meaningless if we use the usual definition of r! However if we adopt the convention that 0! = 1, the formula gives 8!/1 = 8!. Thus we have in general:

Formula 1.2.1

The number of permutations of r objects from n unlike objects = n!/(n−r)!

This is usually written ^nP_r

Example 1.2.2

3 Scotsmen, 5 Welshmen and 4 Irishmen meet to run a mile race. Assuming that all 12 runners have roughly the same best time for this distance, and that all the possible orderings at the end of the race are equally likely to occur find the probability that a Welshman will win, a Scotsman will be second and that Irishmen will be third and fourth.

The number of different ways in which the first 4 places can be filled by these 12 runners is $^{12}P_4$ = 12x11x10x9 = 11,880. If a Welshman wins, he can be any one of 5, if a Scotsman comes second, he can be any one 3, and if Irishmen come third and fourth, they can be chosen in $^4P_2$ = 12 different ways. Thus the required probability

$$= \frac{5 \times 3 \times 12}{12 \times 11 \times 10 \times 9} = 1/66$$

Example 1.2.3

2 teams qualify for the next stage from a 1st round World Cup group of 4. However, all matches end in 0-0 draws, and lots have to be drawn to decide which teams progress. If Wales and Belgium are two of the teams involved, find the probability that
(i) Wales will qualify as winners and Belgium as runners up,
(ii) Wales and Belgium will qualify,
(iii) Wales will qualify as runners up, but Belgium will not qualify at all.

The total number of possible qualifying pairs, taking order into account

$$= {}^4P_2 = 12.$$

(i) Wales as winners, and Belgium as runners up is just one of these possibilities. Thus, the probability of this happening = 1/12
(ii) Wales and Belgium will qualify either with Wales as winners and Belgium as runners up or with Belgium as winners and Wales as runners up. From (i) the probability of either of these events occuring is 1/12. Thus the required probability
$$= 1/12 + 1/12 = 1/6$$
(iii) If Wales are runners up and Belgium do not qualify, then the winner can be any of two teams. Thus the required probability = 2/12 = 1/6.

Self Assessment Questions

1.2.4 Five boys, all of whom are of different heights, are placed in a row. Find the probability that they will be in increasing order of height from left to right.

1.2.5 5 men and 7 women form a committee which has to choose a chairperson, a secretary and a treasurer. Find the probability that both the chairperson and treasurer will be women while the secretary will be a man.

1.2.6 A bag contains 6 red balls, 5 blue balls and 3 yellow balls. 3 balls are removed from the bag without replacement. Find the probability that the first two will be red and the third yellow.

1.2.7 Five different digits are chosen from the set {1,2,3,4,5,6,7,8}. Find the probability that
(i) the first two digits will be even and the final three odd
(ii) the final digit will be 3 greater than the first.

1.2.8 I have eleven books, one of which is a Bible, and another a dictionary. If 5 books are placed on a shelf, find the probability that the left hand book will be the Bible and the right hand book the dictionary.

1.2.9 Both Swansea and Cardiff are through to the last eight of the W.R.U. Cup. Find the probability that when the draw for the quarter finals is made that
(i) the first game out of the bag will result in Swansea being at home to Cardiff,
(ii) Swansea will be at home to Cardiff,
(iii) Swansea will play against Cardiff.

1.3 Combinations

In all the questions done so far, the order in which the objects are arranged has been all important. In many problems however we are not at all interested in the order e.g. if we want to know in how many ways a subcommittee of 3 people can be chosen from a committee of 9, then the composition of the subcommittee is the important factor not the order in which its members are chosen. Arrangements of this kind where the order is irrelevant are called combinations.

Example 1.3.1

When visiting my local music shop, I found 5 cassettes which I would have liked to have bought. They each cost £3, but I had only £9 to spend. How many different combinations of 3 cassettes could I have bought?

Solution 1.3.2 of example 1.3.1

Try and solve this question by letting A,B,C,D,E represent the 5 cassettes, writing down all possible permutations, and then noting how many different permutations give rise to the same combination.

You should have written down $^5P_3 = 5!/2! = 60$ permutations. Further, 6 different permutations always give rise to the same combination. e.g. the permutations
ADE AED DAE DEA EAD EDA all correspond to the single combination {E,A,D} once the order has been ignored, the reason for this being that E,A,D may be arranged amongst themselves in 3! ways = 6 ways. Since this is true for any choice of three letters, we see that the original figure of 60 is 6 times too big, and thus the total number of combinations = 60/6 = 10. This is usually written $^5C_3$ or $\binom{5}{3}$ (the latter notation is used by examination boards).

What we have in effect shown is that
$$^5C_3 = {^5P_3}/3!$$
i.e. that $\quad ^5C_3 = (5!/2!)/3!$
i.e. that $\quad ^5C_3 = 5!/(3! \times 2!)$

Example 1.3.3

Consider question 1.1.11 in which you had to choose 6 books from 10, but this time without concerning yourself with the order in which the books are to be placed on the shelf. In how many ways can this be done?

During the solution of 1.1.11 you should have shown that the number of possible permutations was $^{10}P_4 = 10!/4!$ Let us now denote the books by

A B C D E F G H I J

and consider e.g. the combination consisting of {B, E, G, A, J, C}
The permutations of the books which give rise to this combination correspond to the permutations of the letters B, E, G, A, J, C, of which there are precisely 6! Since this is clearly true for any combination of 6 books, we see that the total number of combinations is given by
$$^{10}C_6 = {^{10}P_6}/6! = (10!/4!)/6!$$
$$= 10!/(6! \times 4!) = 210$$

Exercise 1.3.4

Suggest a general formula for the number of different ways of choosing r objects from n unlike objects if order of choice is to be ignored.

Your result should read as follows

Formula 1.3.5

$$^nC_r = \frac{n!}{r! \times (n-r)!}$$

Example 1.3.6

14 men are named in the squad for a one-day cricket international. In how many ways can 11 players be chosen to play if
(i)   no restriction is made on the choice,
(ii) the captain and the wicketkeeper are certain of their places,
(iii) the 14 men consist of 1 wicketkeeper, 7 batsmen and 6 bowlers and the final 11 must contain the wicketkeeper, 5 batsmen and 5 bowlers.

(i) We must choose any 11 from 14. This can be done in

$$^{14}C_{11} = \frac{14!}{11! \times 3!} = \frac{14 \times 13 \times 12}{3 \times 2 \times 1} = 364 \text{ ways}$$

(ii) Since the captain and wicketkeeper are certain of their places, then the remaining 9 players must be chosen from 12. This can be done in

$$^{12}C_9 = \frac{12!}{9! \times 3!} = \frac{12 \times 11 \times 10}{3 \times 2 \times 1} = 220 \text{ ways}$$

(iii) We must choose 5 batsmen from 7 and 5 bowlers from 6. The batsmen can be chosen in

$$^7C_5 = \frac{7!}{5! \times 2!} = \frac{7 \times 6}{2 \times 1} = 21 \text{ ways}$$

and the bowlers may be chosen in

$$^6C_5 = \frac{6!}{5! \times 1!} = \frac{6}{1} = 6 \text{ ways}$$

Since any one of the 21 choices of batsmen can be included with any one of the 6 choices of bowlers, we see that altogether there are $21 \times 6 = 126$ different ways of choosing the team.

Example 1.3.7

A chess team of 5 is to be chosen at random from 6 boys and 5 girls. Find the probability that the team will contain precisely 3 boys.

The total number of ways in which the team may be chosen

$$^{11}C_5 = \frac{11!}{5! \times 6!} = \frac{11 \times 10 \times 9 \times 8 \times 7}{5 \times 4 \times 3 \times 2 \times 1} = 462 \text{ ways}$$

The number of ways in which we can choose a team containing 3 boys (and hence 2 girls)

$$^6C_3 \times {}^5C_2 = \frac{6!}{3! \times 3!} \times \frac{5!}{2! \times 3!}$$

$$= \frac{6 \times 5 \times 4}{3 \times 2 \times 1} \times \frac{5 \times 4}{2 \times 1}$$

$$= \quad 20 \times 10 \quad = \quad 200$$

Thus the probability that the team will contain precisely 3 boys
$$= 200/462 \quad = \quad 100/231$$

Example 1.3.8

A bag contains 5 red balls and 4 yellow balls. 4 balls are removed at random from the bag and placed to one side. Find the probability that
(i) all are red
(ii) exactly 3 are red
(iii) at least 3 are red
(iv) at least one of the balls is yellow

The total number of ways of choosing 4 balls from 9 $= {}^9C_4 = \dfrac{9!}{4! \times 5!}$

$$= \dfrac{9 \times 8 \times 7 \times 6}{4 \times 3 \times 2 \times 1} = 126$$

(i) 4 red balls can be chosen in ${}^5C_4 = \dfrac{5}{4! \times 1!} = 5/1 = 5$ ways

Thus the probability that all are red $= 5/126$

(ii) 3 red balls (and also 1 yellow ball) may be chosen in

$${}^5C_3 \times {}^4C_1 = \dfrac{5!}{3! \times 2!} \times \dfrac{4!}{1! \times 3!} = \dfrac{5 \times 4}{2 \times 1} \times \dfrac{4}{1} = 40 \text{ ways}$$

Thus, the probability of choosing 3 red balls
$$= \dfrac{40}{126} = \dfrac{20}{63}$$

(iii) P(at least 3 balls are red)
= P(3 balls are red) + P(4 balls are red)
= 40/126 + 5/126
= 45/126
= 5/14

(iv) P(at least one ball is yellow)
= 1− P(all balls are red)
= 1− 5/126   from (i)
= 121/126

Example 1.3.9

4 cards are removed at random from a pack of playing cards. Find the probability that all four will be spades.

The total number of ways of choosing 4 cards from 52 $= {}^{52}C_4$

$$= \dfrac{52!}{4! \times 48!} = \dfrac{52 \times 51 \times 50 \times 49}{4 \times 3 \times 2 \times 1} = 270725$$

If all the cards are to be spades, then we have a choice of 4 from 13. This can be done in

$$^{13}C_4 = \frac{13 \times 12 \times 11 \times 10}{4 \times 3 \times 2 \times 1} = 715 \text{ ways}$$

Thus P(all 4 cards are spades) = 715/270725 = 11/4165

Self-Assessment Questions

1.3.10 In how many ways can a subcommittee of 3 be chosen from a committee of 9?

1.3.11 A 'line' of the football pools consists of choosing 8 games, and to win a first dividend, all 8 games must end in score draws. Alun decides to concentrate on 12 games. How many lines will he need to be sure of a first dividend if exactly 8 of his 12 games actually end in score draws? Comment on the choice of the term 'football permutation' which is usually applied in such situations.

1.3.12 Ten footballers meet to play a five-a-side game. In how many different ways can the two opposing teams be chosen assuming no restrictions on allocating players to the two teams?

1.3.13 (i) In how many ways can we choose 3 letters from the word GERAINT?
(ii) Find the probability that all three will, if randomly chosen, be consonants.

1.3.14 A bag contains 4 white balls and 3 red balls. 2 balls are removed from the bag without replacement. Find the probability that both are red.

1.3.15 (i) 4 people are to be chosen at random from a group of 9 people. In how many ways can this be done?
(ii) Amongst the 9 people there are 2 brothers. Find the probability that both will be chosen.

1.3.16 A box contains 5 red counters and 7 blue counters. 6 counters are removed from the box without replacement. Find the probability that 2 are red and 4 are blue.

1.3.17 5 people are chosen at random from a group of 7 men and 6 women. Find the probability that the number of men chosen will exceed the number of women.

1.3.18 3 digits are chosen at random from the set $\{1,2,3,4,5,6,7\}$. Find the probability that their sum will be odd.

1.3.19 A box contains 3 red beads, 4 white beads and 2 yellow beads. 3 beads are removed at random from the box without replacement. Find the probability that the beads removed are
(i) all white
(ii) all of different colours.

1.3.20 6 cards are dealt from a pack of 52. Find the probability that the 6 cards will
(i) consist of 4 spades and 2 hearts
(ii) consist of 2 spades, 2 hearts and 2 diamonds
(iii) contain precisely 3 diamonds.

MODULE 2

Conditional Probability

2.1 Introduction and development of the concept

The way in which we calculate the probability of the occurence of a particular event depends on precisely what information is at hand about the possible outcomes. For example, if a card is removed from a pack of 52 playing cards, then the probability that it is a king is 4/52 = 1/13. However, if we are given the additional information that the card removed is a picture card, then we would say that its probability of being a king is now 4/12 = 1/3. Consider the following example:

Example 2.1.1

In a sixth form science group of 31 pupils, there are 17 who take Biology and 13 who take Chemistry. There are only 6 students who take neither Biology nor Chemistry.

(i) A pupil is chosen at random from the whole group. Find the probability that he studies Chemistry.

(ii) A pupil is chosen at random from the Biology class. Find the probability that he studies Chemistry.

The information given in the question may be presented in the form of a Venn diagram. In diagram (a) the actual numbers studying each subject are displayed, while diagram (b) gives the corresponding probabilities

(a) $n(B) = 17$
$n(C) = 13$
$n(B \cap C) = 5$

(b) $P(B) = 17/31$
$P(C) = 13/31$
$P(B \cap C) = 5/31$

(i) The probability that a student chosen at random from the whole group studies Chemistry is easily computed direct from the given information. Thus

$$P(C) = \frac{\text{number studying Chemistry}}{\text{number in the whole group}} = \frac{13}{31}$$

However, we can use Venn diagram (a) to give

$$P(C) = \frac{5 + 8}{31} = \frac{13}{31}$$

While diagram (b) gives

$$P(C) = 5/31 + 8/31 = 13/31$$

(ii) If we know that the chosen student studies Biology, then we see from diagram (a) that the sample space is reduced to 17 pupils of whom 5 study Chemistry. Thus we would now say that the probability that the student studies Chemistry is

$$n(B \cap C)/n(B) = 5/17$$

Notice however, that we can achieve the same result by considering the corresponding probabilities from diagram (b) i.e.

$$\frac{P(B \cap C)}{P(B)} = \frac{5/31}{17/31} = \frac{5}{17}$$

What we are in effect doing is finding the probability that a Biology student also studies Chemistry and then expressing this probability not as a fraction of 1, but as a fraction of the probability that the student would study Biology in the first place.

In (ii) we have calculated the probability that a student studies Chemistry **given** that he studies Biology. This is usually written P(C|B) and is an example of **conditional** probability. In this case we have shown that

$$P(C|B) = \frac{P(C \cap B)}{P(B)}$$

and this may be generalised to give the following result:

Formula 2.1.2

If A and B are any two events and P(B) = 0, and P(B) ≠ 0 then

$$P(A|B) = \frac{P(A \cap B)}{P(B)}$$

Example 2.1.3

A fair coin is tossed 5 times. Given that not all tosses come down the same way find the conditional probability that the coin lands heads up precisely twice.

To simplify the notation, let A be the event that not all 5 tosses come down the same way and B the event of getting precisely 2 heads.

Then we need to calculate P(B|A)

The sample space consists of $2^5 = 32$ outcomes, only 2 of which correspond to all 5 tosses coming down the same way (i.e. 5 heads and 5 tails).

Thus n(A) = 32 − 2 = 30

We also note that all these outcomes corresponding to 2 heads (and hence 3 tails) are certainly examples of outcomes in which not all tosses come down the same way. In other words B is a subset of A and hence $A \cap B = B$ thus $n(A \cap B) = n(B) = {}^5C_2 = 10$

Before proceeding any further, you should verify that the above values are correct by writing down all the elements of the sample space and then displaying this information in appropriate Venn diagrams as in example 2.1.1.

Since we are going to use the formula
P(B|A) = P(B ∩ A)/P(A), we must firstly calculate the two probabilities which appear on the right hand side.

$$P(B \cap A) = n(B \cap A)/n(S) = 10/32$$
$$P(A) = n(A)/n(S) = 30/32$$

and hence

$$P(B|A) = \frac{10/32}{30/32} = 10/30 = 1/3$$

Example 2.1.4

2 cards are dealt face down form a shuffled pack of 52 playing cards. One card is turned over and is seen to be a club. Find the conditional probability that the other card is also a club.

Let A be the event that the card turned over is a club, and B the event that the card not turned over is a club. Then we must calculated P(B|A)

P(A ∩ B) = the probability that both cards are clubs

$$= \frac{{}^{13}C_2}{{}^{52}C_2} = \frac{(13 \times 12)/(2 \times 1)}{(52/51)/(2 \times 1)} = \frac{13 \times 12}{52 \times 51} = \frac{1}{17}$$

P(A) = 13/52 = 1/4

Thus from formula 2.1.2, we have

$$P(B|A) = \frac{P(A \cap B)}{P(A)} = \frac{1/17}{1/4} = \frac{4}{17}$$

We can easily verify this result is true by noting that once we know that one card is a club, then 12 of the remaining 51 cards must be clubs, and thus the probability that the other card dealt is also a club

$$= 12/51 = 4/17$$

Example 2.1.5

3/4 of the pupils in a particular class have dark hair, while 2/5 of the class have both dark hair and brown eyes. A pupil who is chosen at random is found to have dark hair. Find the probability that he also has brown eyes.

Let A be the event that a pupil has dark hair and B the event that a pupil has brown eyes. We must calculate $P(B|A)$
We are given $P(A) = 3/4$, $P(A \cap B) = 2/5$

Thus $P(B|A) = P(A \cap B) / P(A) = \dfrac{2/5}{3/4} = \dfrac{8}{15}$

Example 2.1.6

A card is chosen at random from a pack of 35 cards numbered 1,2,3,..., 35. Given that the number on the card is even, find the conditional probability that it is divisible by 4.

Let A be the event that the number on the card is even and B the event that the number on the card is divisible by 4.
We must calculate $P(B|A)$
Clearly $P(A) = 17/35$
Also $A \cap B = B$ since all numbers which are divisible by 4 must also be even, and thus
$P(A \cap B) = P(B) = 8/35$

Thus $P(B|A) = \dfrac{P(A \cap B)}{P(A)} = (8/35) / (17/35) = 8/17$

Example 2.1.7

A bag contains 4 white counters, 2 red counters and 3 yellow counters. 3 counters are removed at random from the bag without replacement.
(i)   Find the probability that none of the counters is white
(ii) Given that at least one of the counters is white, find the probability that all 3 are white.

Let A be the event that none of the counters is white,
B the event that at least one of the counters is white, and
C the event that all three counters are white.

(i)   Then $P(A) = {}^5C_3 / {}^9C_3 = 10/84 = 5/42$
(ii) We must calculate $P(C|B)$
      $P(B) = 1 - P(A) = 37/42$
      Since C is a subset of B, then $C \cap B = C$, and hence
      $P(C \cap B) = P(C) = {}^4C_3 / {}^9C_3 = \dfrac{4}{84} = \dfrac{1}{21}$

      Thus $P(C|B) = \dfrac{P(C \cap B)}{P(B)} = \dfrac{1/21}{37/42} = \dfrac{2}{37}$

Before proceeding to the Self Assessment Questions, we shall again remind the student that what we have in effect done in all the above examples is to restrict the sample space. We shall return again to this interpretation of conditional probability in Module 4 where it can in many cases lead to an intuitive understanding of certain probabilities which would otherwise have to be calculated either by using Probability Tree Diagrams, the Law of Total Probability or Bayes' Theorem.

Self Assessment Questions

2.1.8 All the pupils in the top stream in the 5th form of a particular school sit both English and Mathematics at GCSE. None of these pupils failed both subjects, 3/4 of them passed in Mathematics and 2/3 of them passed in English. (i) Draw a Venn diagram to find the fraction of pupils who passed in both subjects, (ii) A pupil chosen at random was found to have passed in Mathematics. Find the conditional probability that he also passed in English.

2.1.9 The probability that Ian Rush will score in any game is 2/5. The probability that he will score a hat-trick in any game is 1/30. Given that Ian Rush scored last Saturday, find the probability that he scored a hat-trick.

2.1.10 A card is chosen at random from a pack of 150 cards numbered 1 to 150. Given that the number on it is divisible by 3 find the conditional probability that it is also divisible by 5.

2.1.11 A domino is removed at random from an ordinary set (i) Find the probability that the sum of the dots on the domino is odd. (ii) Given that the sum of the dots on the domino is even, find the conditional probability that it is a 'double'.

2.1.12 Two fair dice are thrown, one red and one blue. Find the conditional probability that
(i) the sum of the numbers on the two dice is 8 given that the number on the red die is odd
(ii) the number on the red die is odd given that the sum of the numbers on the two dice is 8.

2.1.13 A bag contains 7 red and 3 white balls. Two balls are removed without replacement. Given that at least one ball is red, find the conditional probability that both balls are red.

2.1.14 Two cards are chosen at random without replacement from a pack of 8 cards numbered from 1 to 8. Find the conditional probability that
(i) the sum of the two numbers is odd given that their product is even.
(ii) the product of the two numbers is even and given that their sum is even.

2.1.15 A committee consists of 10 men and 5 women. A subcommittee of 4 is to be formed and must contain at least one woman. Find the conditional probability that the other 3 members will all be men.

2.2 The multiplication formulae

The result $P(B|A) = P(A \cap B)/P(A)$ may be crossmultiplied to give

Formula 2.2.1

$$P(A \cap B) = P(A).P(B|A)$$

This form of the equation may be derived directly as in the following example:

Example 2.2.2

A bag contains 8 blue balls and 5 red balls. If 2 balls are removed without replacement, find the probability that both are blue.

This problem can be very quickly solved by using the methods of Module 1, the required probability being

$$\frac{{}^8C_2}{{}^{13}C_2} = \frac{(8\times 7)/(2\times 1)}{(13\times 12)/(2\times 1)} = \frac{8\times 7}{13\times 12} = \frac{14}{39}$$

It can also be approached by interpreting probability as the proportion of successes in a large number of trials (Empirical probability). Assume that the above experiment is repeated many times. Then we would expect the first ball drawn to be blue in approximately 8/13 of the trials, since 8 of the 13 balls in the bag are blue. Furthermore the expected proportion of these trials (i.e. those in which the first ball is blue) in which the second ball drawn is also blue is 7/12, since 7 of the 12 balls now remaining are blue. Therefore, we would expect the overall proportion of trials in which both balls are blue to be 7/12 of 8/13

$$= 7/12 \times 8/13 = 14/39 \text{ as above.}$$

Using B_1 to denote the event that the first ball drawn is blue and B_2 the event that the second ball drawn is blue, and noting that
$$P(B_2) = 8/13 \text{ and } P(B_2|B_1) = 7/12$$
we deduce that
$$P(B_1 \cap B_2) = P(B_1).P(B_2|B_1)$$
which corresponds to the general formula 2.2.1.

Example 2.2.3

It rains in Blaenau Ffestiniog on 1/2 of the days of the year. On 1/3 of those days on which it rains, it will have started raining before midday. Find the probability that it will rain next Sunday morning.

Let A denote the event that it will rain next Sunday and B the event that it will rain next Sunday morning. We must find P(B).
We are given that $P(A) = 1/2$ and that $P(B|A) = 1/3$
Thus $P(A \cap B) = P(A).P(B|A) = 1/2 \times 1/3 = 1/6$
However B is a subset of A and thus $A \cap B = B$
Hence $P(B) = P(A \cap B) = 1/6$

Example 2.2.4

Two cards are chosen at random without replacement from the 13 spades which have been removed from a pack of playing cards. Find the probability that one card is a picture card and the other is not.

We must consider two possibilities

(a) The first card drawn is a picture card and the second is not. Let A_1 be the event that the first card drawn is a picture card and A_2' the event that the second card drawn is not a picture card
Then $P(A_1) = 3/13 \quad P(A_2'|A_1) = 10/12$
and thus $P(A_1 \cap A_2') = 3/13 \times 10/12 = 5/26$

(b) The first card is not a picture card whereas the second is. Using the above notation
Let A_1' be the event that the first card drawn is not a picture card
and A_2 the event that the second card drawn is a picture card
Then $P(A_1') = 10/13$ and $P(A_2|A_1') = 3/12$
and thus $P(A_1' \cap A_2) = 10/13 \times 3/12 = 5/26$

Since events (a) and (b) are mutually exclusive we see that the required probability is equal to $5/26 + 5/26 = 5/13$

You should now check by using the methods of Module 1.

Example 2.2.5

Two cards are removed from a pack of playing cards. Find the probability that the first card is an Ace and the second a club.

This question is complicated by the fact that the ace removed may or may not be the Ace of clubs. We therefore split up the problem to take this into account.
Let A_1 be the event that the first card removed is the Ace of clubs,
A_2 be the event that the first card removed is an Ace other than the Ace of Clubs, and C the event that the second card removed is a club.
Then $P(A_1) = 1/52$ and $P(C|A_1) = 12/51$
Thus $P(A_1 \cap C) = 1/52 \times 12/51 = 12/2652$
Also $P(A_2) = 3/52$ and $P(C|A_2) = 13/51$
Thus $P(A_2 \cap C) = 3/52 \times 13/51 = 39/2652$

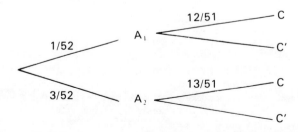

The events $A_1 \cap C$ and $A_2 \cap C$ are mutually exclusive, and hence the required probability is
$P(A_1 \cap C) + P(A_2 \cap C)$
$= 12/2652 + 39/2652 = 51/2652 = 1/52$

The result given in formula 2.2.1 can be extended to three or more events. If A,B and C are any three events, we can calculate $P(A \cap B \cap C)$ as follows.
To begin with, we consider $A \cap B$ as one event, and thus

$$P(A \cap B \cap C) = P([A \cap B] \cap C)$$
$$= P(A \cap B).P(C|A \cap B)$$
But $P(A \cap B) = P(A).P(B|A)$ and thus we have

Formula 2.2.6

$$P(A \cap B \cap C) = P(A).P(B|A).P(C|A \cap B)$$

In other words, the probability of A,B and C is the probability of A × the probability of B given A × the probability of C given both A and B.

Example 2.2.7

Lots are drawn to decide the order in which 10 children take turns on the school computer. If Gwen, Einir and Dafydd are 3 of these children, find the probability that Dafydd will be first, Gwen will be second and Einir will be third.

Let D be the event that Dafydd will be first, G that Gwen will be second and E that Einir will be third
$P(D \cap G \cap E) = P(D).P(G|D).P(E|D \cap G)$
$P(D) = 1/10 \quad P(G|D) = 1/9 \quad P(E|D \cap G) = 1/8$
and thus
$P(D \cap G \cap E) = 1/10 \times 1/9 \times 1/8 = 1/720$

Exercise 2.2.8

(i) Prove that if A,B,C and D are four events then
$P(A \cap B \cap C \cap D) = P(A).P(B|A).P(C|A \cap B).P(D|A \cap B \cap C)$
(ii) What is the corresponding result for n events $A_1, A_2, A_3 \ldots A_n$

Examples 2.2.9

A bag contains 3 red balls and 2 white balls. If the five balls are removed from the bag without replacement, find the probability that the second and fourth balls removed are white.

We exhibit two methods for solving this problem

Method 1

If the second and fourth balls are white then the first, third and fifth must be red. Let
R_1 be the event that the first ball is red
W_2 the event that the second ball is white
R_3 the event that the third ball is red
W_4 the event that the fourth ball is white and
R_5 the event that the fifth ball is red

Then $P(R_1 \cap W_2 \cap R_3 \cap W_4 \cap R_5) =$
$P(R_1).P(W_2|R_1).P(R_3|R_1 \cap W_2).P(W_4|R_1 \cap W_2 \cap R_3).P(R_5|R_1 \cap W_2 \cap R_3 \cap W_4)$
But $P(R_1) = 3/5$
$P(W_2|R_1) = 2/4$
$P(R_3|R_1 \cap W_2) = 2/3$
$P(W_4|R_1 \cap W_2 \cap R_3) = 1/2$
$P(R_5|R_1 \cap W_2 \cap R_3 \cap W_4) = 1$
Thus
$P(R_1 \cap W_2 \cap R_3 \cap W_4 \cap R_5) = 3/5 \times 2/4 \times 2/3 \times 1/2 \times 1$
$= 1/10$

Method 2

Since any one of the 5 balls is equally as likely as any other one to be removed as second, then using the notation of method 1 we have $P(W_2) = 2/5$. Furthermore if the only information we have is that the second ball was white, the probability that the fourth was white is 1/4 i.e.

$$P(W_4|W_2) = 1/4$$

Thus $P(W_2 \cap W_4) = P(W_2).P(W_4|W_2)$
$= 2/5 \times 1/4$
$= 1/10$

Example 2.2.10

11 cards which are numbered from 1 to 11 are shuffled and placed face down in a pile. The top 5 cards are then turned over. Find the probability that
(i)   all 5 cards are odd
(ii) the first three are odd and the next two are even
(iii) 3 of the cards are odd and 2 are even.

By using the results derived in 2.2.8(ii) for 5 events we have
(i)   P(all 5 cards odd) = 6/11 x 5/10 x 4/9 x 3/8 x 2/7 = 1/77
(ii) P(1st 3 odd and next 2 even) = 6/11 x 5/10 x 4/9 x 5/8 x 4/7 = 10/231
(iii) Let us now consider another arrangement which corresponds to 3 odd cards and 2 even cards e.g. where the 2nd and 4th cards are the only ones with even numbers. Then using obvious notation, we can calculate the probability of this arrangement occurring from
$P(O_1 \cap O_3 \cap O_5 \cap E_2 \cap E_4) =$
$= P(O_1) \times P(O_3 | O_1) \times P(O_5 | O_1 \cap O_3) \times P(E_2 | O_1 \cap O_3 \cap O_5) \times P(E_4 | O_1 \cap O_3 \cap O_5 \cap E_2)$
$= $ 6/11 x 5/10 x 4/9 x 5/8 x 4/7 = 10/231

This type of argument will clearly work for any arrangement corresponding to 3 odd numbers and 2 even numbers, and since there are 5!/(3!x2!) = 10 such arrangements, the required probability is

10 x 10/231 = 100/231

Note 2.2.11

In the above solution, the order in which we applied our formula to the events in question was different to the order in which the events actually occurred. Arguing like this can often lead to quick solutions as was the case in Method 2 of example 2.2.9. It also explains why the two probabilities involved in example 2.2.9 turned out to be the same.

If you are still unconvinced of the validity of this argument you should return to the arrangement considerd in part (iii) above and apply the formula to the events in the order in which they occurred.

Self assessment questions

2.2.12 One half of the pupils in a particular school have blond hair, and 2/3 of these children also have blue eyes. Find the probability that a pupil chosen at random will have both blond hair and blue eyes.

2.2.13 In my class at school, there are 16 boys and 12 girls. The names are placed on the register in alphabetical order irrespective of sex. Find the probability that the first name on the register is that of a boy and the second that of a girl.

2.2.14 Whenever Vivian Richards goes to the crease, there is a 1 in 4 chance that he will pass 50. If he does so there is a 2/5 chance that he will carry on to score a century. Find the probability that he will score a century in his next innings.

2.2.15 When playing table tennis, I seem to grow in confidence whenever I win a point, so that the probability of my winning the next point is 3/5. On the other hand, if I lose a point, then I only have a 1 in 3 chance of winning the next one. I have just won a point. Find the probability that I will win the one after next.

2.2.16 A bag contains 5 red balls numbered 1 to 5, 4 blue balls numbered 6 to 9, and 3 yellow balls numbered 10 − 12. Two balls are removed without replacement. Find the probability that the first ball is red and the second ball has an even number on it.

2.2.17 A box contains 5 yellow balls, 3 red balls and 2 white balls. 3 balls are removed without replacement. Find the probability that
(i)   the first is red, the second yellow and the third is white
(ii) the first is white, the second red and the third red
(iii) the three balls removed will consist of two red balls and one yellow ball.

2.2.18 All 7 counters are removed one by one from a box which originally contained 2 white counters and 5 yellow counters. Find the probability that the two white counters are removed consecutively.

2.2.19 7 people are chosen at random from a group of 9 men and 7 women. Find the probability that 4 will be men and 3 will be women.

2.3 The relationship with module 1

Many of the problems encountered in 2.2 are similar in nature to questions which were solved by using permutations and combinations in Module 1. We shall now underline the fact that some questions may be tackled in more than one way by giving alternative solutions to examples 1.2.2 and 1.3.9. It should however be noted that the methods of Module 1 are valid only for equally likely outcomes whereas the following method is valid in all situations.

Example 2.3.1

Let us consider again example 1.2.2
Let W_1 be the event that a Welshman will win, S_2 that a Scotsman will come second, I_3 that an Irishman will come third, and I_4 that an Irishman will come fourth.
Then $P(W_1 \cap S_2 \cap I_3 \cap I_4)$ = $P(W_1) \times P(S_2|W_1) \times P(I_3|W_1 \cap S_2) \times P(I_4|W_1 \cap S_2 \cap I_3)$

$$= 5/12 \times 3/11 \times 4/10 \times 3/9$$
$$= 1/66$$

Example 2.3.2

Consider example 1.3.9
Then using obvious notation,
$P(S_1 \cap S_2 \cap S_3 \cap S_4)$ = $P(S_1) \times P(S_2|S_1) \times P(S_3|S_1 \cap S_2) \times P(S_4|S_1 \cap S_2 \cap S_3)$

$$= 13/52 \times 12/51 \times 11/50 \times 10/49 = 11/4165$$

If the question were to ask further for the probability of getting 2 hearts, 1 diamond and 1 club, then we would proceed as follows -
$P(H_1 \cap H_2 \cap D_3 \cap C_4)$ = $P(H_1) \times P(H_2|H_1) \times P(D_3|H_1 \cap H_2) \times P(C_4|H_1 \cap H_2 \cap D_3)$

$$= 13/52 \times 12/51 \times 13/50 \times 13/49$$
$$= 169/41650$$

However, this is only one of $4!/2! = 12$ arrangements which correspond to 2 hearts, 1 diamond and 1 club, and by arguing as in 2.2.10, we see that the required probability
$$= 12 \times 169/41650 = 1014/20825$$

Self assessment questions

Questions 2.3.3. → 2.3.19
Use the multiplication formulae for conditional probability to solve 1.2.3, 1.2.4, 1.2.5, 1.2.6, 1.2.7, 1.2.8, 1.2.9, 1.3.7, 1.3.8, 1.3.13 (ii), 1.3.14, 1.5.15(ii), 1.3.16, 1.3.17, 1.3.18, 1.3.19, 1.3.20

After completing the self assessment questions, you should now be aware of the fact that the conditional probability formulae can provide quick and elegant solutions to many problems. However, examples such as 2.3.15, 2.3.16 and 2.3.19 are still probably better tackled by using combinations, and thus it is important that you are able to understand and master both methods of solution.

2.4 A useful result

If X and Y are any two events, then one way of calculating the probability that both X and Y occur is to use the formula
$$P(X \cap Y) = P(X).P(Y|X)$$
However, since $Y \cap X = X \cap Y$, we could also use
$$P(Y \cap X) = P(Y).P(X|Y)$$
By comparing the right hand sides of these two equations, we see that
$$P(X).P(Y|X) = P(Y).P(X|Y) \text{ or equivalently}$$

Formula 2.4.1

$$P(Y|X) = \frac{P(Y).P(X|Y)}{P(X)}$$

The following example gives an application of this formula.

Example 2.4.2

Sometimes I walk to school, otherwise I go on my bike. The probability that I will walk to school on any given day is 1/3, but if I do decide to walk, there is a 1/4 chance that I will be late. My form master tells me that on average, I am late on 1 day in 10. Given that I was late yesterday, find the conditional probability that I walked to school.

Let W be the event 'I walked to school yesterday', and L the event 'I was late yesterday'

Then we are given $P(W) = 1/3$, $P(L|W) = 1/4$, $P(L) = 1/10$

We need to calculate $P(W|L)$.
From the formula,

$$P(W|L) = \frac{P(W).P(L|W)}{P(L)}$$

$$= \frac{1/3 \times 1/4}{1/10} = \frac{10}{12} = \frac{5}{6}$$

Self Assessment Questions

2.4.3 Mark Hughes and Ian Rush form a formidable striking partnership for Wales. The probability that Hughes will score in any international is 2/5, while the corresponding probability for Rush is 3/10. In those games in which Hughes scores, there is a 1/8 chance that Rush will score as well. Given that Rush scored for Wales last night, find the probability that Hughes also scored.

2.4.4 In our GCSE class last year three-quarters of us passed in English and 3/5 passed in History. 2/3 of those who passed in English also passed in History. What proportion of those who passed in History also passed in English?

2.4.5 I am not very good at forecasting the weather. If it rains on any given day, there is only a 1/2 chance that I will have taken my umbrella with me; on the other hand, if I have taken my umbrella with me, the probability that it will rain is only 2/3. In our town it rains on 2/5 of the days of the year. How often do I take my umbrella with me?

2.4.6 Two events A and B are such that $P(A|B) = 0$. Find (i) $P(B|A)$ and (ii) $P(A \cap B)$. How would you describe the relationship between the events A and B?

TUTOR-ASSESSED QUESTIONS 1 (Modules 1 & 2)

1. (a) The letters {a,a,a,b,n,n} are arranged in random order. Find the probability that they will spell the word banana.

   (b) If any three letters of the same set are arranged in random order, find the probability that they spell the word ban.

2. (a) A delegation of 4 persons is to be chosen at random from 5 married couples. Find the probability that the delegation will contain:
      (i) 2 married couples,
      (ii) at least one man and at least one woman,
      (iii) no married couple.

   (b) A well shuffled pack of playing cards is dealt out to four players, each receiving 13 cards. Show that the probability that a particular player receives four aces is approximately 0.0026.

3. A bag contains 9 white balls, 7 blue balls and 4 green balls. If three balls are removed at random without replacement, find the probability that
   (i) all three balls are of different colours,
   (ii) two balls are white and one is green,
   (iii) the first ball drawn is white and the other two are not white.

4. (a) Two numbers are chosen at random from a table of random numbers containing the numbers 0,1,2,...,11. Find the probability that
      (i) the sum of the two numbers is greater than 11, given that the first number is 5,
      (ii) the second number is 4, given that the sum of the two numbers is greater than 8,
      (iii) the first number is 5, given that the difference between the two numbers is 5.

   (b) At least one goal is scored in 90% of all league matches. The probability that a league match will end in a draw is 2/9. Given that Swansea's last game was drawn, find the conditional probability that it was a score draw.

5. 5 cards are chosen at random from an ordinary pack of playing cards. Find the probability that
   (i) The first will be a heart, the second a diamond, both the third and fourth clubs, and the fifth a heart.
   (ii) Two will be hearts and three will be spades.
   (iii) All will be diamonds.
   (iv) The third card will be the only club.

6. (a) On any given day, there is a 1/4 probability that it will rain at 8.00 a.m. and a 1/6 probability that it will rain at 4.00 p.m. On 1/3 of those days when it does rain at 4.00 p.m. it was already raining at 8.00 a.m. How good an indicator as to whether it will be raining at 4.00 p.m. is the weather at 8.00 a.m.?

   (b) In 1861, 90% of the population of a particular village were illiterate. Of those who were literate, 40% were innumerate and of those who were numerate, 70% were illiterate. What proportion of the inhabitants of the village were innumerate?

MODULE 3

Independence

3.1 Intuitive definition of independence

Two events are said to be **independent** if the occurrence or non-occurrence of one event in no way affects the probability of the occurrence or non-occurrence of the other. Consider the following examples.

Example 3.1.1.

An unbiased die is thrown twice. Let A be the event that the first throw is a 6, B the event that the second throw is a 3 and C the event that the sum of the two throws is at least 8.

Then A and B are independent, since the outcome of the first throw does not affect the outcome of the second throw. On the other hand A and C are not independent since getting a 6 on the first throw clearly improves the probability that the total will be at least 8 as any score greater than 1 on the second throw will give the desired result.

Example 3.1.2

A bag contains 4 white and 6 red balls. A ball is removed and its colour noted, and then without replacing the first ball, a second ball is removed. If A is the event that the first ball removed is white, and B the event that the second ball removed is red, then A and B are not independent since the probability that the second ball removed is red given that the first ball removed was also red is 5/9 i.e. $P(B|A') = 5/9$ whereas the probability that the second ball is red given that the first ball removed is white is 6/9 i.e. $P(B|A) = 6/9$. If however the first ball is replaced before the second ball is removed, then A and B are independent since in this case $P(B) = 6/10$ irrespective of which colour ball was first removed.

Self-Assessment Questions

Decide whether or not the following pairs of events are independent.

3.1.3 Two fair coins are tossed. A is the event that the first coin is a head, and B the event that the second coin is a tail.

3.1.4 Two cards are removed from a pack of playing cards (a) with replacement (b) without replacement; X is the event that the first card is a spade, and Y the event that the second card is a spade.

3.1.5 A fair coin is tossed 10 times. N is the event that all the first nine tosses are heads, and L that the 10th toss is a head.

3.1.6 Liverpool are playing against Everton. D is the event that the game is drawn, and E the event that an even number of goals are scored.

3.1.7 On the same day as Liverpool are playing Everton, Tottenham are playing West Ham. L is the event that Liverpool beat Everton, and W the event that West Ham beat Tottenham.

3.1.8 An unbiased die is thrown once. X is the event that the number on the die is divisible by 4, and Y the event that the number on the die is greater than 4.

3.1.9 An unbiased die is thrown once. A is the event that the number on the die is divisible by 3 and B the event that the number on the die is greater than 3.

3.2 Mathematical definition of independence

In some of the self-assessment questions above (3.1.8 and 3.1.9 perhaps), it is not intuitively obvious whether the two given events are independent or not. In order to deal with such situations we must develop a mathematical definition of independence, and we do this as follows.

If we return to our original definition, we see that for A and B to be independent, the probability of B occurring must not be affected by A. In mathematical terms, this is equivalent to saying that

3.2.1 $\quad P(B|A) = P(B)$

Since $P(B|A) = \dfrac{P(A \cap B)}{P(A)}$

we must also have $P(B) = \dfrac{P(A \cap B)}{P(A)}$

or on cross-multiplying

3.2.2. $\quad P(A).P(B) = P(A \cap B)$

Furthermore, since $P(B|A) = \dfrac{P(B).P(A|B)}{P(A)}$

then
$$P(B) = \dfrac{P(B).P(A|B)}{P(A)}$$

and hence

3.2.3 $\quad P(A|B) = P(A)$

You should not be too surprised to see that we have derived 3.2.3 from 3.2.1. All we have in fact done is to prove that if B is independent of A, then A is independent of B. It is also important to realise that if we wish to check if two events A and B are independent, we need only consider whether any **one** of equations 3.2.1, 3.2.2 or 3.2.3 is satisfied.

Example 3.2.4

Consider question 3.1.8 where $P(X) = 1/6$, but $P(X|Y) = 0$, and hence from 3.2.3, X and Y are not independent.

However, if we now consider question 3.1.9, we see that
$$P(A) = 1/3, \ P(B) = 1/2 \text{ and } P(A \cap B) = 1/6$$
Since $P(A).P(B) = P(A \cap B)$, then A and B are independent from 3.2.2.

Self-Assessment Questions

3.2.5 A card is removed at random from a pack of 52 playing cards. Let A, B, C be the events that the card chosen is a heart, a queen and a black card respectively. Investigate whether the following pairs of events are independent or not:

(i) A and B
(ii) B and C
(iii) A and C

3.2.6 A red die and a blue die are thrown simultaneously. Let A be the event that the score on the red die is 2, and B the event that the sum of the scores on the two dice is 8. Are A and B independent?

3.2.7 In a street of 120 houses, 24 houses take the Mirror and 40 houses take the Sun. 64 houses take neither of these papers. Let S be the event that a house in this street takes the Sun, and M the event that a house in this street takes the Mirror. Are the events S and M independent?

3.2.8 Two events A and B are such that $P(A) = 1/4$, $P(A \cup B) = 2/3$, $P(B) = 1/2$. Are A and B independent?

3.2.9 The local weather station has gathered the following information from readings taken in recent years. The probability that it will rain on any given day is 1/3. There is high wind on

1/4 of these days on which it rains, but on 1/2 of the days there is neither rain nor high wind. Does the fact that there is a high wind make it more or less likely to rain?

3.3 Problems involving 2 independent events

Example 3.3.1

A fair die is thrown twice. Find the probability that:
(i) the first throw gives an odd number and the second throw gives a 5,
(ii) at least one 3 is thrown.

The outcome of the second throw is clearly independent of that of the first throw.
(i) Let A be the event of getting an odd number on the first throw, and B the event of getting a 6 on the second throw.
Since A and B are independent, then
$P(A \cap B) = P(A).P(B)$
But $P(A) = 3/6 = 1/2$ and $P(B) = 1/6$
Thus $P(A \cap B) = 1/2 \times 1/6 = 1/12$

(ii) This part is more easily answered by first considering the probability of not getting a 3, as the desired event is the complement of this.

Let A be the event of not getting a 3 on the first throw, and B the event of not getting a 3 on the second throw.

Then as A and B are independent
$P(A \cap B) = P(A).P(B)$
But $P(A) = P(B) = 5/6$, and thus $P(A \cap B) = 5/6 \times 5/6 = 25/36$
The probability which we have calculated is that of not getting a 3 on either throw. Thus the probability of getting at least one 3 is $1 - (25/36) = 11/36$

Example 3.3.2

2/5 of the population of a particular town wear glasses, while 2/3 of the population are 5'5'' or taller. Assuming that these two properties are independent, find the probability that a person chosen at random will be

(i)   at least 5'5'' tall and also wear glasses
(ii) under 5'5'' tall and does not wear glasses
(iii) either at least 5'5'' tall or wears glasses.

Let A be the event that a person wears glasses, and B the event that a person is at least 5'5'' tall.
(i) We require $P(A \cap B)$. From independence, $P(A \cap B) = P(A).P(B) = 2/5 \times 2/3 = 4/15$.
(ii) We require $P(A' \cap B')$. Since A and B are independent, then so are A' and B'.
Thus $P(A' \cap B') = P(A').P(B') = 3/4 \times 1/3 = 1/5$
(iii) We require $P(A \cup B)$
Now $P(A \cup B) = P(A) + P(B) - P(A \cap B) = 2/5 + 2/3 - 4/15 = 12/15 = 4/5$.

Self-Assessment Questions

3.3.3 A fair coin is tossed and an unbiased die is thrown. Find the probability of getting a head on the coin and a number divisible by 3 on the die.

3.3.4 A pack consists of ten cards numbered from 1 to 10. A card is removed from the pack and its value noted. It is then replaced, the pack shuffled, and a second card is removed. Find the probability that

(i)   the number on each card is 3
(ii) the number on the first card is less than 5 and the number on the second is divisible by 3
(iii) the number on at least one card is divisible by 5.

3.3.5 The probability that I am late for work on any given day is 1/25. Find the probability that on two consecutive mornings

(i) I am late twice
(ii) I am not late at all
(iii) I am late once.

3.3.6 A machine contains two components A and B which work independently of one another, and will only operate properly if both A and B are functioning. The probability that A will fail is 0.01, while the corresponding probability for B is 0.05. Find the probability that the machine is working properly.

3.3.7 Alun and Gwyn visit their local Youth Club independently of one another although Alun is there far more often than Gwyn. The probability that both will be there on the same night is 1/8, while the probability that neither will be there is 5/24. Find the probability that Alun will be there tonight.

3.3.8 The probability that I will have chips for dinner is 1/5, and independently, the probability that I will have ice-cream for pudding is 1/6. Find the probability that tomorrow I will either have chips or ice-cream, but not both.

3.4 Independence of 3 events

We can generalize our intuitive idea of independence to 3 events A, B and C. They will be independent if the occurrence of any one event or any pair of the events does not affect the probability of the occurrence of any other event. Thus, in particular, the events must be **pairwise** independent i.e.

$$P(A \cap B) = P(A).P(B)$$
$$P(A \cap C) = P(A).P(C)$$
$$P(B \cap C) = P(B).P(C)$$

However, by considering the equation $P(A \cap B \cap C) = P(A).P(B|A).P(C|A \cap B)$ and noting that our definition above implies that $P(B|A) = P(B)$ and $P(C|A \cap B) = P(C)$ we reach the following definition for the **total** independence of 3 events.

Definition 3.4.1

The events A, B and C are totally independent if
$$P(A \cap B) = P(A).P(B)$$
$$P(A \cap C) = P(A).P(C)$$
$$P(B \cap C) = P(B).P(C)$$
$$P(A \cap B \cap C) = P(A).P(B).P(C)$$

Example 3.4.2

Three fair coins are tossed.
Let A be the event that the first coin is a head
B the event that the second coin is a tail
and C the event that the third coin is a head.

Intuitively, A, B and C ought to be independent events. We can check whether this is the case by constructing the sample space, and then using the mathematical definiton of total independence. In obvious notation S = {HHH, HHT, HTH, HTT, THH, THT, TTH, TTT} and thus
$P(A) = P(B) = P(C) = 4/8 = 1/2$
$A \cap B = \{HTH, HTT\}$ and thus $P(A \cap B) = 2/8 = 1/4$
$A \cap C = \{HTH, HHH\}$ and thus $P(A \cap C) = 2/8 = 1/4$
$B \cap C = \{HTH, TTH\}$ and thus $P(B \cap C) = 2/8 = 1/4$
$A \cap B \cap C = \{HTH\}$ and thus $P(A \cap B \cap C) = 1/8$
From the above information we see that
$P(A \cap B) = P(A).P(B)$

$P(A \cap C) = P(A).P(C)$
$P(B \cap C) = P(B).P(C)$
$P(A \cap B \cap C) = P(A).P(B).P(C)$
Thus A, B and C are totally independent events.

Self-Assessment Questions

It is recommended that you construct the sample space in each of the following questions.

3.4.3 A red die and a blue die are thrown simultaneuosly. Let A be the event that the score on the red die is divisible by 3, B the event that the score on the red die is a prime number and C the event that the score on the blue die is greater than 2. Prove that A, B and C are totally independent.

3.4.4 A fair coin is tossed three times. Let M be the event that the outcomes on the first and second tosses are different, N the event that the outcomes on the second and third tosses are different and Q the event that the outcome of the first toss is a head. Prove that M, N and Q are totally independent.

3.4.5 A fair coin is tossed twice. Let A be the event that the outcome of the first toss is a head, B the event that the outcome of the second toss is a head, and C the event that the outcomes of the two tosses are different. Show that A, B and C are pairwise independent but not totally independent.

3.4.6 A fair coin is tossed three times. Let S be the event that the outcome of the first toss is a head, T the event that the outcome of the third toss is a tail and U the event that the outcome of at least 2 tosses are heads. Show that $P(S \cap T \cap U) = P(S).P(T).P(U)$ but that S, T and U are not totally independent.

3.5 Problems involving 3 or more independent events

It is clear that our intuitive definition of independence may be extended to any finite number of events, but a complete mathematical definition becomes more and more complicated as the number of events increases. It suffices to note that a **necessary** condition for n events $A_1, A_2, \ldots A_n$ to be totally independent is that

3.5.1 $P(A_1 \cap A_2 \cap A_3 \cap \ldots A_n) = P(A_1).P(A_2).P(A_4) \ldots P(A_n)$

i.e. given that $\{A_1, A_2, A_3, \ldots A_n\}$ are independent, then the above results must hold. On the other hand, it is important to realize that this is not a **sufficient** condition. It is still possible for the above equation to be satisfied without $\{A_1, A_2, \ldots A_n\}$ being totally independent (See question 3.4.6 for the case n = 3).

Example 3.5.2

Three cards are drawn at random from an ordinary pack of playing cards, each card being replaced before the next one is removed. Find the probability that
(i) All three cards will be spades
(ii) The first card is any Queen, while the second and third are hearts
(iii) All three will be face cards
(iv) There will be at least one diamond amongst the three cards.

Since each card is always replaced before the next one is removed, the outcomes of the three draws are totally independent.

(i) Let S_1 be the event that the first card is a spade,
  S_2 the event that the second card is a spade,
  and S_3 the event that the third card is a spade.
  The $P(S_1) = P(S_2) = P(S_3) = 13/52 = 1/4$
  From independence
  $P(S_1 \cap S_2 \cap S_3) = P(S_2).P(S_2).P(S_3)$
  $= 1/4 \times 1/4 \times 1/4$
  $= 1/64$

(ii) Let Q_1, H_2, H_3 represent the given events (using obvious notation)
Then $P(Q_1) = 4/52 = 1/13$ $P(H_2) = P(H_3) = 13/52 = 1/4$
Thus $P(Q_1 \cap H_2 \cap H_3) = P(Q_1).P(H_2).P(H_3)$
$= 1/13 \times 1/4 \times 1/4 = 1/208$

(iii) Let F_1, F_2, F_3 represent the given events
Then $P(F_1) = P(F_2) = P(F_3) = 12/52 = 3/13$
and hence $P(F_1 \cap F_2 \cap F_3) = P(F_1) \times P(F_2) \times P(F_3)$
$= 3/13 \times 3/13 \times 3/13$
$= 27/2197$

(iv) To answer this part of the question, we first of all calculate the probability that there will be no diamonds chosen. We use D_1', D_2', D_3' to denote the events that the chosen cards are not diamonds. Then D_1', D_2', D_3' are totally independent and $P(D_1') = P(D_2') = P(D_3')$
$= 39/52 = 3/4$
Thus $P(D_1' \cap D_2' \cap D_3') = P(D_1') \times P(D_2') \times P(D_3')$
$= 3/4 \times 3/4 \times 3/4$
$= 27/64$
Thus the probability that at least one diamond is chosen is
$1 - (27/64) = 37/64$

Example 3.5.3

The probability that a particular rifleman hits a moving target with any one shot is 1/3. Assuming that all hits are independent events

(a) Find the probability that
  (i) he misses the target with each of his first 8 shots
  (ii) the first hit is the third shot.

(b) How many shots are necessary before he has a greater than 99 in 100 chance of having hit the target at least once?

Let H_r be the event that the rifleman hits the r th target. Then the H_r are totally independent, and P(Hr) always equals 1/3.

(a) (i) We must calculate $P(H_1' \cap H_2' \cap \ldots H_8')$
Now $P(H_r') = 2/3$ for all r, and thus from independence
$P(H_1' \cap H_2' \cap \ldots H_8') = (2/3)^8 = 256/6561$

(ii) We must calculate $P(H_1' \cap H_2' \cap H_3)$
From independence $P(H_1' \cap H_2' \cap H_3) = P(H_1') \times P(H_2') \times P(H_3)$
$= 2/3 \times 2/3 \times 1/3$
$= 4/27$

(b) The final part is equivalent to finding how many shots are necessary before the probability of his missing all the targets becomes less than 1 in 100. If n is the required number, then arguing as in (a) (i) we must have $(2/3)^n < 1/100$
Taking logs this is equivalent to
$n \log_{10}(2/3) < \log_{10}(1/100)$
and since $\log_{10}(2/3)$ is negative, we must have
$$n > \frac{\log_{10}(1/100)}{\log_{10}(2/3)}$$
i.e. $n > 11.36$
Thus at least 12 shots are necessary.

Example 3.5.4

I either walk to school or I go on my bike. The probability that I will decide to walk on any given day

is always 3/5. Find the probability that in any given week of 5 days, I will
(i) not go on my bike at all
(ii) cycle at least once
(iii) cycle precisely once
(iv) cycle at most twice.

Assume that the way in which I travel to school on any given day is independent of how I go on any other day.

Let B_1, B_2, B_3, B_4, B_5 represent the events that I cycle to school on the 1st, 2nd, 3rd, 4th and 5th days of the given week.

(i) We must calculate $P(B_1' \cap B_2' \cap B_3' \cap B_4' \cap B_5')$, which from independence equals

$P(B_1') \times P(B_2') \times P(B_3') \times P(B_4') \times P(B_5') = (3/5)^5 = 243/3125$

(ii) P(cycling at least once) = 1 − P(not cycling at all)
= 1 − (243/3125)
= 2882/3125

(iii) Let us for the moment calculate the probability that I only cycle to school on Monday. Then we must calculate $P(B_1 \cap B_2' \cap B_3' \cap B_4' \cap B_5')$
= $P(B_1) \times P(B_2') \times P(B_3') \times P(B_4') \times P(B_5')$
= $2/5 \times (3/5)^4 = 162/3125$

However, the probability that I only cycle to school on Tuesday will also be 162/3125, and similarly for Wednesday, Thursday and Friday. Since these events are mutually exclusive then the probability that I will cycle to school precisely once = 5 × (162/3125) = 162/625.

(iv) We shall first calculate the probability that I will cycle to school precisely twice. If these two days are to be Monday and Tuesday, then we must calculate
$P(B_1 \cap B_2 \cap B_3' \cap B_4' \cap B_5')$
= $P(B_1) \times P(B_2) \times P(B_3') \times P(B_4') \times P(B_5')$
= $(2/5)^2 \times (3/5)^3$
= 108/3125

But this would also be the probability of my cycling to school on just Tuesday and Friday, and similarly for any other two chosen days out of 5. Thus, the probability of my cycling to school on precisely two days

= $^5C_2 \times 108/3125$
= 10 × 108/3125
= 216/625

The event that I cycle to school at most twice is the disjoint union of the events {not cycling at all}, {cycling precisely once}, and {cycling precisely twice}.
These events are mutually exclusive, and hence the required probability equals
243/3125 + 162/625 + 216/625 = 2133/3125

Self-assessment questions

3.5.5 In a set of 10 multiple choice questions, there are 3 possible answers to each question. A candidate decides to choose his answers to each question at random. Find the probability that he gets
(i) all his answers incorrect
(ii) at least one answer correct
(iii) precisely one answer correct.

3.5.6 Three unbiased coins are tossed. Find the probability of getting
(i) 3 tails
(ii) no tails
(iii) 1 tail
(iv) 2 tails
Check your answers by constructing the sample space.

3.5.7 4 seeds are placed in a seed box. The probability that any individual seed will germinate is 2/5.

Find the probability that
(i) all 4 seeds will germinate
(ii) no seeds germinate
(iii) precisely 2 seeds germinate.
Assume that the germination of any seed is independent of the germination of any other.

3.5.8 A and B play a game by taking turns to throw a die, the first to throw a 4 being the winner. A goes first.
(a) Find the probability that
(i) A will win on his first throw
(ii) B will win on his first throw
(iii) A will win on his third throw
(iv) B will win on his fifth throw.

(b) Find the least value of n such that the probability that the game is still in progress after each has had n turns is less than 1 in 1000.

3.5.9 A bag contains three red balls and one white ball. A, B and C play a game in which they take turns to remove a ball at random from the bag, note the colour, and then replace it, the first to remove the white ball being the winner. A goes first, then B and then C.
Find the probabilities that
(i) A wins on his first turn
(ii) A wins on his second turn
(iii) A wins on his third turn
(iv) A wins on or before his third turn
(v) A wins on or before his 10th turn
(vi) A wins.

MODULE 4

Probability tree diagrams, total probability and Bayes' Theorem

4.1 Probability Tree Diagrams

Whenever an experiment may be considered as consisting of a number of different stages, the probabilities involved may be displayed on a probability tree diagram. Consider the following problem

Example 4.1.1

Three bags A, B and C each contain 10 balls. A contains 4 white, 3 red and 3 yellow balls, B contains 2 white, 6 red and 2 yellow balls, while C contains 4 white, 1 red and 5 yellow balls. An unbiased die is thrown. If the score on the die is 1, a ball is chosen at random from the bag A, if the score is 2 or 3, a ball is chosen at random from bag B, otherwise a ball is chosen at random from bag C. Find the probability that the ball chosen is (i) white (ii) red (iii) yellow

The two stages of this question are clearly defined. At the first stage, a die is thrown resulting in a choice being made of one of the three bags. The second stage involves removing a ball, which has one of three colours from the chosen bag. Hence there are 3 x 3 = 9 possible events which we must consider, and these may be displayed as follows

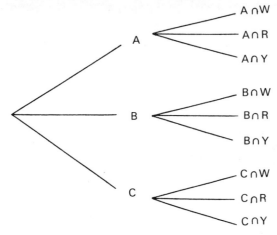

where e.g. B∩Y means firstly choosing bag B and then removing a yellow ball from it. This is the basis of the probability tree diagram.

Let us now calculate the individual probabilities involved with different parts of the diagram. We first derive P(A), P(B) and P(C). A will be the chosen bag only if the score on the die is 1 and thus P(A) = 1/6. Similarly, P(B) = 2/6 and P(C) = 3/6. These values should be entered on the diagram as shown below

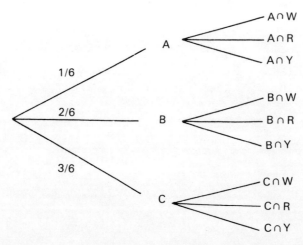

We eventually want to calculate the probabilities of the events which appear on the right hand side of the diagram. But to calculate e.g. P(A∩W) we need to use the formula

P(A∩W) = P(A).(W|A)

and thus first of all we must calculate P(W|A). This is however easily done, since 4 of the 10 balls in A are white, and thus P(W|A) = 4/10. Similarly P(R|A) = 3/10 and P(Y|C) = 5/10. These values together with the remaining conditional probabilities may now be entered on the diagram.

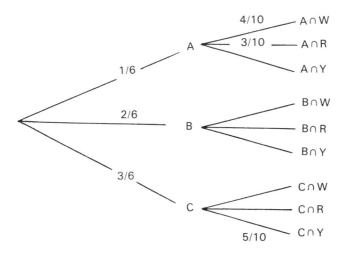

The use of the multiplication formula given above now enables us to compute the required probabilities. Thus e.g.

P(A∩W) = P(A).P(W|A) = 1/6 × 4/10 = 4/60
P(A∩R) = P(A).P(R|A) = 1/6 × 3/10 = 3/60
P(C∩Y) = P(C).P(Y|C) = 3/6 × 5/10 = 15/60

These values are entered on the right hand side of the diagram below. You should finish the probability tree diagram yourself before looking at the completed diagram.

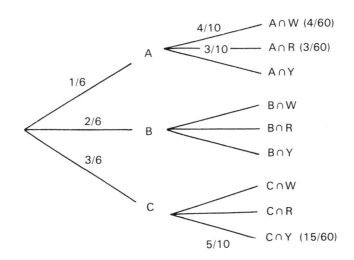

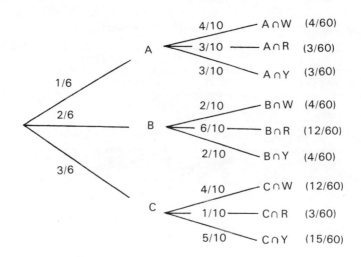

Let us now return to the original problem.

(i) We see from our diagrams that there are three events which correspond to choosing a white ball i.e. A∩W, B∩W, and C∩W. Since these events are mutually exclusive then

$$P(W) = P(A \cap W) + P(B \cap W) + P(C \cap W)$$
$$= 4/60 + 4/60 + 12/60$$
$$= 20/60 = 1/3$$

(ii) Similarly

$$P(R) = P(A \cap R) + P(B \cap R) + P(C \cap R)$$
$$= 3/60 + 12/60 + 3/60$$
$$= 18/60 = 3/10$$

(iii)

$$P(Y) = P(A \cap Y) + P(B \cap Y) + P(C \cap Y)$$
$$= 3/60 + 4/60 + 15/60$$
$$= 22/60 = 11/30$$

Note that since choosing a white ball, choosing a red ball and choosing a yellow ball are mutually exclusive events, the sum of the probabilities calculated in (i) (ii) and (iii) must equal 1.

In problems such as the one above, there are many advantages in drawing a tree diagram. First of all, it gives the student a complete picture of the probabilities of all the possible outcomes of an experiment. Furthermore, since using the multiplication rule actually corresponds to multiplying along the branches of the diagram, the calculation can be done in a simple and systematic way, and the probabilities can be displayed in a form which may be easily understood and interpreted. Of course, in practice, the probability tree need only be drawn once, but it is always advisable to draw a large and clear diagram.

Note that at every stage of the above experiment

(i) All possible outcomes are considered, i.e. the events are exhaustive
(ii) No two events have any outcome in common, i.e. the events are mutually exclusive
Consequently
(a) The sum of the probabilities along the branches leaving any point is always 1
(b) The sum of the probabilities on the right hand side of the diagram is 1

These facts should always be checked when drawing a probability tree diagram, and this is why the values of the probabilities as displayed on the branches are best left uncancelled.

Example 4.1.2
A bag contains 6 red and 4 white balls. 3 balls are removed at random from the bag without replacement. Find the probability that two are red and one is white.

This time, the experiment may be thought of as consisting of three stages, each are corresponding to the removal of a ball. Using obvious notation, the probability tree diagram is constructed as follows:

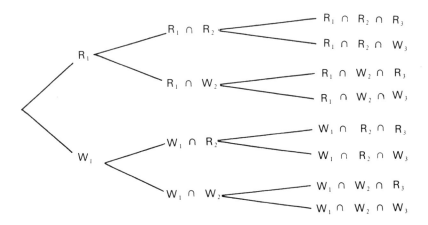

At the first stage

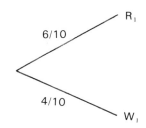

At the second stage

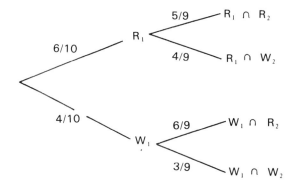

Remember — the probabilities on each branch are conditional probabilities. Thus the 5/9 on the top branch represents $P(R_2|R_1)$, and has been calculated from the knowledge that if a red ball has been removed at the first stage, then 5 of the remaining 9 balls are red.

At the third stage

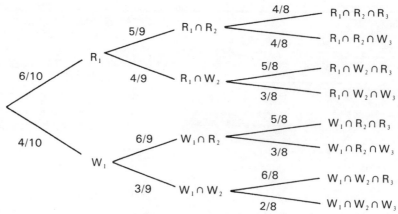

The 4/8 on the top branch represents $P(R_3 | R_1 \cap R_2)$, which takes this value since 4 of the remaining 8 balls are red.

The probabilities of the eight events which appear on the right hand side of the diagram can now be calculated by multiplying along the appropriate branches from left to right. Thus e.g.

$P(R_1 \cap R_2 \cap R_3) = P(R_1) \times P(R_2 | R_1) \times P(R_3 | R_1 \cap R_2)$
$= 6/10 \times 5/9 \times 4/8$
$= 120/720$

The completed diagram is given below:

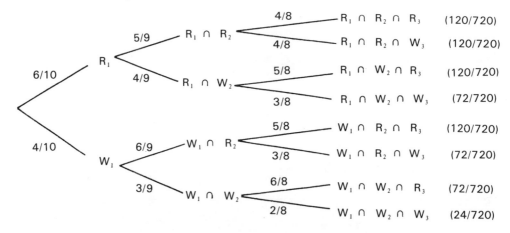

If we now return to the original question, we can see that

P(two red balls and one white ball)
$= P(R_1 \cap R_2 \cap W_3) + P(R_1 \cap W_2 \cap R_3) + P(W_1 \cap R_2 \cap R_3)$
$= 120/720 + 120/720 + 120/720$
$= 360/720 = 1/2$

Exercise 4.1.3

Check this result by using the methods of Module 1 and Module 2.

Example 4.1.4

I travel to work by car 1/3 of the time, on my motor bike 1/4 of the time. I cycle 3/8 of the time, otherwise I jog. The respective probabilities that I will be late for work are 1/10 if I go by car, 1/15 if

I go by motor bike, 1/20 if I cycle and 1/5 if I jog.
(i) Find the probability that I will be late for work next Tuesday.
(ii) Given that I was late for work yesterday, find the conditional probability that I cycled.

This problem differs from 4.1.1 and 4.1.2 in that it may not be immediately obvious that it may be divided up into different stages. However, after a little thought, we see that we can think of a first stage involving the different methods of travelling to work, and a second involving either being late or being on time. Thus, the probability tree diagram is constructed as follows, using obvious notation.

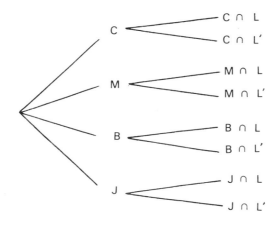

1st stage

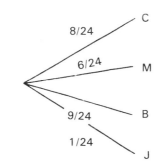

Note (i) $P(J) = 1 - 1/3 - 1/4 - 3/8 = 1/24$
(ii) All probabilities have been put over a common denominator.

2nd stage

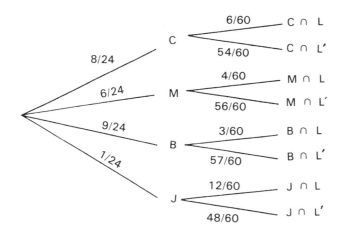

All probabilities have again been placed over a common denominator. The fact that e.g. P(L'|C) = 54/60 follows immediately from the fact that there is a 9/10 chance of my being late for work if I travel by car.

Finally

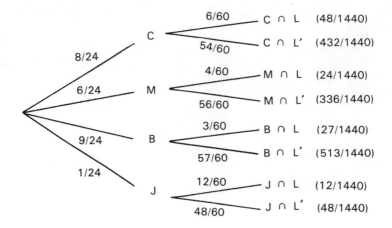

Thus
(i) P(L) = P(C∩L) + P(M∩L) + P(B∩L) + P(J∩L)
   = 48/1440 + 24/1440 + 27/1440 + 12/1440
   = 111/1440 = 37/480

(ii) $P(B|L) = \dfrac{P(B \cap L)}{P(L)} = \dfrac{27/1440}{111/1440} = \dfrac{27}{111} = \dfrac{9}{37}$

It is important to realize that we can also achieve this result directly from the probability tree diagram by restricting the sample space. Since it is given that I was late for work yesterday, the restricted sample space becomes the union of the four events

C∩L, M∩L, B∩L, and J∩L

whose combined probability is

44/1440 + 24/1440 + 27/1440 + 12/1440

Of the above events, the one which corresponds to my cycling to work is B∩L which has probability 27/1440.

Thus intuitively

$P(B|L) = \dfrac{27/1440}{48/1440 + 24/1440 + 27/1440 + 12/1440}$

$= \dfrac{27/1440}{111/1440} = \dfrac{27}{111} = \dfrac{9}{37}$

The fact that all denominators are the same enables us to simplify the calculation further to

$\dfrac{27}{48 + 24 + 27 + 12} = \dfrac{27}{111} = \dfrac{9}{37}$

Self-assessment questions

Solve the following questions by using probability tree diagrams.

4.1.5 I always try to arrive in school on time, but this depends on whether or not I sleep late. If I wake up on time, there is only a 1 in 10 chance of my being late for school. On the other hand, if I sleep late, then there is a 2/3 chance of my arriving late. If the probability of my waking up on time on any given day is 4/5, find the probability that I will be late for school tomorrow.

4.1.6 A biased coin is such that the probability of getting a head is three times that of getting a tail. If this coin is tossed three times, find the probability of getting precisely one head amongst the three tosses.

4.1.7 Two thirds of the people attending a basic German language course have no previous knowledge of the language. It is certainly an advantage to have already studied the language because then the probability of passing the examination at the end of the course is 3/4 whereas otherwise it is only 1/2. Find the probability that a person chosen at random from those attending the course will pass the examination.

4.1.8 Alun takes Welsh, Mathematics and French at A-level. The probability that he will pass in Welsh is 0.6, in Mathematics 0.8, and in French 0.5, and these three probabilities are totally independent of one another. In order to follow the course of his choice after leaving school, he needs to pass at least two A-levels. Find the probability that he will succeed in doing this.

4.1.9 A bag contains 8 blue and 4 red marbles. Three marbles are removed without replacement. Find the probability that at least 2 are blue.

4.1.10 Bag A contains 4 white and 3 red balls, and bag B contains 2 white and 5 red balls. A coin is tossed and if the outcome is a head, a ball is removed at random from A, otherwise a ball is removed at random from B. The coin is however biased, being twice as likely to give a head as it is to give a tail. (i) Find the probability that the ball chosen will be red. (ii) Given that the ball chosen is red, find the conditional probability that it came from A.

4.1.11 Alun, Elfyn and Dafydd help at home by doing the washing up after supper. Alun does it 1/2 of the time, Elfyn does it on average on 1 night in 3, and Dafydd does it the rest of the time. The probability that something will be broken when Alun washes up is 1/40, whereas Elfyn and Dafydd are less careful, and their respective probabilities of breaking something are 1/30 and 1/20. (i) Find the probability that something will be broken tonight. (ii) Given that nothing was broken last night find the probability that Elfyn did the washing up.

4.2 The Law of Total Probability

During the solution of Example 4.1.1 we noted that at all stages of the probability tree diagram, the events considered were both exhaustive and mutually exclusive. We shall now consider these concepts more carefully while giving an alternative method for solving Example 4.1.4.

If we consider the first stage of our problem, then the events denoted by C, M, B and J are both exhaustive, since they cover all possible outcomes and also mutually exclusive since only one of these methods can be used for getting to work on any given day.

If a Venn diagram were now drawn to represent these events, the sample space would be partitioned as shown below:

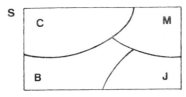

Now consider the event L superimposed on the above Venn diagram. We must take into account the possibility of L intersecting each of the above subsets (indeed in this case we know that this is so), and the Venn diagram would look like:

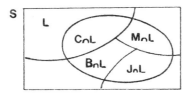

Thus L itself is partitioned into disjoint subsets $C \cap L$, $M \cap L$, $B \cap L$, and $J \cap L$, and hence
$$P(L) = P(C \cap L) + P(M \cap L) + P(B \cap L) + P(J \cap L)$$
But we can take this one step further, since
$$P(C \cap L) = P(C).P(L|C)$$
$$P(M \cap L) = P(M).P(L|M) \text{ etc}$$
and thus we have:
$$P(L) = P(C).P(L|C) + P(M).P(L|M) + P(B).P(L|B) + P(J).P(L|J)$$
Since all these quantities are either given in the question or are easily calculated, we see that
$$P(L) = 1/3 \times 1/10 + 1/4 \times 1/15 + 3/8 \times 1/20 + 1/24 \times 1/5$$
$$= 37/480 \text{ as above.}$$

This gives us an alternative method for solving the problem without drawing the associated probability tree diagram. It is an example of the use of the Law of Total Probability, which we now formally define.

Definition 4.2.1 (The Law of Total Probability)

Let $\{A_1, A_2, ..., A_n\}$ be a set of mutually exclusive and exhaustive events.
Let B be any other event. Then
$$P(B) = P(A_1).P(B|A_1) + P(A_2).P(B|A_2) + \ldots + P(A_n).P(B|A_n)$$

Example 4.2.2

The probability that a person wears glasses seems to vary according to age. 1 person in 5 whose age is less than 21 wears glasses, while the corresponding proportions for the age groups 21—30, 31—60, and 61 and over are 1 in 8, 2 in 5 and 7 in 10. In our town 25% of the population are under 21, 15% are between 21 and 30, 40% are between 31 and 60 and 20% are 61 or over.

(i) Find the probability that a person chosen at random from our town will wear glasses.
(ii) Given that a person chosen at random wears glasses, find the probability that he/she is under 21.

To begin with, we must decide on our notation, and to help us to do this, we look carefully at the question. We are first of all asked to calculate the probability that a person chosen at random will wear glasses, and this will be the event denoted by B. On the other hand, four age groups partition the population of the town into disjoint subsets, and the corresponding events will be our A_i ($i = 1, 2, 3, 4$). Thus if

B is the event that a person wears glasses
A_1 the event that a person is under 21 years old
A_2 the event that a person is between 21 and 30 years old
A_3 the event that a person is between 31 and 60 years old
A_4 the event that a person is 61 or over.

Then $\{A_1, A_2, A_3, A_4\}$ form a set of mutually exclusive and exhaustive events, and thus we may use the Law of Total Probability.

(i) We are given
$P(A_1) = 25/100$ $P(B|A_1) = 1/5 = 8/40$
$P(A_2) = 15/100$ $P(B|A_2) = 1/8 = 5/40$
$P(A_3) = 40/100$ $P(B|A_3) = 2/5 = 16/40$
$P(A_4) = 20/100$ $P(B|A_4) = 7/10 = 28/40$
Thus
$$P(B) = P(A_1).P(B|A_1) + P(A_2).P(B|A_2) + P(A_3).P(B|A_3) + P(A_4).P(B|A_4)$$
$$= 25/100 \times 8/40 + 15/100 \times 5/40 + 40/100 \times 16/40 + 20/100 \times 28/40$$
$$= 200/4000 + 75/4000 + 640/4000 + 560/4000$$
$$= 1475/4000$$
$$= 59/160$$

(ii) We must calculate $P(A_1|B)$
Now
$$P(A_1|B) = \frac{P(A_1 \cap B)}{P(B)} = \frac{P(A_1).P(B|A_1)}{P(B)}$$

$$= \frac{200/4000}{1475/4000} = 200/1475 = 8/59$$

Part (ii) may be more clearly explained by drawing a Venn diagram. Since we must calculate $P(A_1 | B)$, we restrict our sample space to B, and the probabilities corresponding to the four subsets of B are entered on the diagram as shown:

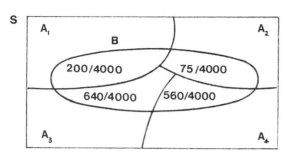

Thus $P(A_1 | B)$

$$= \frac{200/4000}{200/4000 + 75/4000 + 640/4000 + 560/4000}$$

$$= \frac{200}{200 + 75 + 640 + 560}$$

$$= 200/1475 = 8/59$$

Self-assessment questions

4.2.3, 4.2.4, 4.2.5, 4.2.6.
Repeat exercises 4.1.5, 4.1.7, 4.1.10 and 4.1.11 without drawing probability tree diagrams.

4.2.7 A particular make of light bulb is produced in one of 3 factories X, Y and Z, whose outputs are in the ratios 3:2:1 respectively. The probability that a bulb from factory X will last for more than 2000 hours is 1/5, while the corresponding probabilities for bulbs produced in factories Y and Z are 1/6 and 1/8 respectively. I have just bought one of these bulbs from the local supermarket. Find the probability that it will last for more than 2000 hours.

4.2.8 I travel to work by route A, B or C. I go by route A on average 1 day in 4. I go by route B on average 1 day in 3, otherwise I go by route C. If I go by route A, the probability that I will be late is 1/10, while the corresponding probabilities for routes B and C are 1/5 and 1/20 respectively. Find the probability that I was late for work last Thursday.

4.2.9 In a recent local opinion poll, electors were asked (a) which party they would vote for in the next election (b) whether they were in favour of Sunday trading. 30% said they would vote Conservative, 35% said they would vote Labour, 25% said they would vote Alliance and 10% said they would vote Plaid Cymru. 40% of the Conservative voters were in favour of Sunday trading, while the corresponding percentages for Labour, Alliance and Plaid Cymru were 50%, 70% and 30%.
(i) What was the overall percentage of those people interviewed who were in favour of Sunday trading?
(ii) Given that a particular person interviewed stated that he was in favour of Sunday trading, find the conditional probability that he intended voting Alliance.

4.3 Bayes' Theorem

In the previous two sections on Probability Tree Diagrams and the Law of Total Probability, some of the examples and exercises involved the calculation of conditional probabilities. The questions involved are Example 4.1.4, self-assessment questions 4.1.10, 4.1.11, Example 4.2.2. and self-assessment questions 4.2.5, 4.2.6 and 4.2.9. We shall now look more closely at some of these conditional probabilities with a view to obtaining a general formula. First of all we shall consider Example 4.1.4, and you are advised to try this again for yourself before proceeding any further.

In this question, we were concerned with calculating $P(B|L)$ i.e. the probability of my having travelled by bike given that I was late for work, and to do this, we used the usual formula

$$P(B|L) = \frac{P(B \cap L)}{P(L)}$$

However $P(L)$ had itself been calculated by adding together the probabilities of the mutually exclusive events $C \cap L$, $B \cap L$, $M \cap L$, and $J \cap L$, and thus we may write

$$P(B|L) = \frac{P(B \cap L)}{P(C \cap L) + P(B \cap L) + P(M \cap L) + P(J \cap L)}$$

We also saw that we could easily explain this result by looking at the Probability Tree Diagram and restricting the sample space to the events which correspond to the occurence of event L. $P(B|L)$ is then calculated by considering $P(B \cap L)$ as a fraction of the probability of the restricted sample space.

The above result can however be developed one step further since
$$P(B \cap L) = P(B).P(L|B)$$
and by using similar expressions for $P(C \cap L)$, $P(M \cap L)$ and $P(J \cap L)$ we have

$$P(B|L) = \frac{P(B).P(L|B)}{P(C).P(L|C) + P(B).P(L|B) + P(M).P(L|M) + P(J).P(L|J)}$$

We now of course recognize the denominator as being equal to $P(L)$ by the Law of Total Probability. We also note that the expression for $P(B|L)$ contains precisely those probabilities given to us in the question, and hence this result gives us a very quick method of calculating the required probability. Furthermore, there is clearly a pattern in the formula in that when calculating $P(B|L)$, the numerator becomes $P(B).P(L|B)$, while the denominator is the sum of similar expressions involving the set of exhaustive and mutually exclusive events $\{C,B,M,J\}$

You are now recommended to read through Example 4.2.2. This time we needed to calculate $P(A_1|B)$ and this was done from the formula

$$P(A_1|B) = \frac{P(A_1).P(B|A_1)}{P(B)}$$

We then inserted the value for $P(B)$ which had been calculated from the Law of Total Probability, and thus the formula may be rewritten as

$$P(A_1|B) = \frac{P(A_1).P(B|A_1)}{P(A_1).P(B|A_1) + P(A_2).P(B|A_2) + P(A_3).P(B|A_3) + P(A_4).P(B|A_4)}$$

It is now important to realize that the comments made above with respect to the formula for $P(B|L)$ are also valid in this case.

Exercise 4.3.1

Derive similar formulae for the conditional probabilities calculated in questions 4.1.10, 4.1.11 and 4.2.9. By looking at your results, conjecture what form a general formula might take, and what conditions the events involved must satisfy. This is given in 4.3.2 below.

Theorem 4.3.2 (Bayes' Theorem)

Let $\{A_1, A_2, \ldots A_n\}$ be a set of mutually exclusive and exhaustive events, and let A_r be any member of this set. If B is any other event then

$$P(A_r | B) = \frac{P(A_r).P(B|A_r)}{P(A_1).P(B|A_1) + P(A_2).P(B|A_2) + \ldots + P(A_n).P(B|A_n)}$$

A simple proof of this result may be given as follows

$$P(A_r | B) = \frac{P(A_r).P(B|A_r)}{P(B)} \quad \text{(by formula 2.4.1)}$$

$$= \frac{P(A_r).P(B|A_r)}{P(A_1).P(B|A_1) + P(A_2).P(B|A_2) + \ldots + P(A_n).P(B|A_n)}$$

(by the Law of Total Probability 4.2.1)

Example 4.3.3

Bag X contains 7 white balls and 3 red balls, while bag Y contains 3 white balls and 5 red balls. After choosing a bag, I remove a ball from its contents at random. However, I am twice as likely to choose bag X as bag Y. Given that the ball chosen was white, find the conditional probability that the bag chosen was Y.

We need to find
P(bag chosen was Y | ball chosen was white)
Thus, if we compare with the theorem as written above,
Let B be the event that the ball chosen was white,
A_1 that the bag chosen was X
and A_2 that the bag chosen was Y.
Then we have

$P(A_1) = 2/3$ $P(B|A_1) = 7/20 = 28/40$
$P(A_2) = 1/3$ $P(B|A_2) = 3/8 = 15/40$

By Bayes' Theorem

$$P(A_2|B) = \frac{P(A_2).P(B|A_2)}{P(A_1).P(B|A_1) + P(A_2).P(B|A_2)}$$

$$= \frac{1/3 \times 15/40}{2/3 \times 28/40 + 1/3 \times 15/40}$$

$$= \frac{15}{56 + 15} = \frac{15}{71}$$

Example 4.3.4

I play as striker for my local football team. Over the past few years, we have won 60% of our games, drawn 25% and lost the rest. When we win, there is a 1 in 3 chance that I score, if we lose then there is only a 1 in 5 chance that I will have scored. On average, I score in 1 in 6 of our drawn games. Given that I scored in our last game, find the conditional probability that we won.

We need to find P(we won | I scored). If we compare with Bayes' Theorem as written above, then if B is the event that I scored,
A_1 that we won,
A_2 that we lost,
A_3 that we drew

then we have

$P(A_1) = 12/20$ $P(B|A_1) = 1/3 = 10/30$
$P(A_2) = 3/20$ $P(B|A_2) = 1/5 = 6/30$
$P(A_3) = 5/20$ $P(B|A_3) = 1/6 = 5/30$

By Bayes' Theorem

$$P(A_1|B) = \frac{P(A_1).P(B|A_1)}{P(A_1).P(B|A_1) + P(A_2).P(B|A_2) + P(A_3).P(B|A_3)}$$

$$= \frac{12/20 \times 10/30}{12/20 \times 10/30 + 3/20 \times 6/30 + 5/20 \times 5/30}$$

$$= \frac{120}{120 + 18 + 25} = \frac{120}{163}$$

Before proceeding to the self-assessment questions, we make the following comments about the use of Bayes' Theorem.

1. The most difficult part can often be deciding how to choose the notation. This can best be done by writing down what we have to calculate, so that we immediately know which events to denote by B and A_r. (It does not really matter which subscript we choose at this stage. For simplicity's sake we can always use A_1). The remainder of the A_i must then be chosen so that $(A_1, A_2, \ldots, A_n)$ is a set of mutually exclusive and exhaustive events.

2. It is usually worthwhile expressing all the $P(A_i)$ as fractions with the same denominator and similarly for all the $P(B|A_i)$, since this makes the final calculation far simpler.

3. Bayes' Theorem gives us a quick method for calculating conditional probabilities for which we would otherwise need to draw Probability Tree Diagrams, and thus its use can be advantageous under examination conditions. The price we pay is that we no longer have a complete picture of the probabilities of all those events which form the sample space. If you do have difficulty in using Bayes' Theorem, you can always look for help by constructing the associated Probability Tree Diagram.

Self-assessment questions

4.3.5 Three bags X, Y and Z each contain a mixture of red and white balls. X contains 4 red and 3 white balls, Y contains 5 red and 4 white balls, and Z contains 2 red and 3 white balls. A bag is chosen at random, and a ball is removed at random from this bag. Given that the ball removed was found to be white, find the conditional probability that the bag chosen was Z.

4.3.6 A particular electrical item is produced in one of three factories A, B and C. A makes 30% of all such items produced, while B and C make 60% and 10% respectively. 2% of A's output are defective, while the corresponding percentages for B and C are 1% and 3%. I have just bought one of those items and found that it does not work. Find the conditional probability that it was made in factory C.

4.3.7 One person in 10,000 suffers from a certain rare disease. A test to discover the presence of the disease gives a positive reaction for 95% of the people suffering from the disease, but also for 1% of people not suffering from the disease. If the test gives a positive reaction to a randomly selected person, find the conditional probability that he/she has the disease.

4.3.8 I try to go out jogging as often as possible. On average, I run every other day on weekdays, but on both Saturdays and Sundays, there is a probability of 3/4 that I will go out. Given that I ran today, find the conditional probability that it is either a Saturday or a Sunday.

4.3.9 Bag A contains 6 white balls and 3 red balls, while bag B contains 5 white balls and 4 red balls. A ball is taken at random from A and placed in B. A ball is now removed from B and is found to be white. Find the conditional probability that the ball transferred from A to B was red.

4.3.10 10% of players playing county cricket throw left handed. 60% of left handed throwers also bat left handed, but amongst right handed throwers only 30% are left handed batsmen. Find the probability that a right handed batsman also throws right handed.

4.3.11 Today I sat a multiple choice examination in a subject where on average I know the correct answer to one question in three. For each question, 5 possible answers were given, and I decided that when I did not know the correct answer, I would choose any one of the given answers at random. The answer I gave to question 1 was correct. Find the probability that I guessed it.

4.3.12 Apples from a particular orchard are classified into one of four categories according to size. 20% go into category A, 40% into category B, 30% into category C and the rest go into category D. The probability that an apple chosen from category A will be bruised is 0.08, while the corresponding probabilities for apples from categories B, C and D are 0.06, 0.04 and 0.03. Given that a particular apple picked from this orchard turns out to be bruised, find the conditional probability that it was placed in category B.

TUTOR-ASSESSED QUESTIONS 2 (Modules 3 & 4)

1. (a) A and B are independent events such that:
   (i) P(A|B) = 1/3 and
   (ii) P(B|A) = 3/8
   Find P(A and B)

   (b) The following observations were made when a red die and a blue die, neither of which is unbaised, were rolled simultaneously a large number of times:
   (i) the red die showed 6 far more often than the blue die,
   (ii) in 12% of the trials, both dice showed 6,
   (iii) in 32% of the trials, neither die showed 6.
   Use this information to find the probability that on any one throw, the blue die will show a 6. How many times is the red die more likely to show a 6 than the blue die?

2. Two rifleman A and B have independent probabilities 5/8 and 1/8 respectively of hitting a target.
   (i) If each fires twice, what is the probability that the target will be hit by at least one shot?
   (ii) If B fires once only, how many times must A shoot in order to make the probability of hitting the target at least once greater than 0.999?

3. Two snooker players, Alun and Bryn decide to play a match against one another where the winner will be the first to win 3 frames. No frame can end in a tie, and Alun's probability of winning any such frame is always 0.6. Find the probability that:
   (i) Alun wins the first three frames,
   (ii) the match finishes after 3 frames,
   (iii) Alun wins the match after the fourth frame,
   (iv) the match finishes after 4 frames,
   (v) Alun wins the match.

4. (a) The probability that a golfer hits the ball onto a green if it is windy as he strikes the ball is 0.3, and the corresponding probability if it is not windy as he strikes the ball is 0.6. The probability that the wind will blow as he strikes the ball is 0.2. Draw a probability tree diagram to represent this information, and use it to find the probability that:
   (i) he hits the ball onto the green,
   (ii) it was not windy, given that he does not hit the ball onto the green.

   (b) Events A and B are such that P(A) = 1/6, P(B|A) = 1/5 and P(B'|A') = 3/4. By drawing a probability tree diagram, find
   (i) P(A and not B)
   (ii) P(B)
   (iii) P(A|B)
   (iv) P(A or B)

5. (a) Three boxes contain coloured balls, box 1 contains 7 red and 3 blue, box 2 contains 4 red and 6 blue, while box 3 contains 2 red and 8 blue. A box is chosen and then a ball is chosen at random from this box. If the probabilities of choosing the boxes are 5/8, 1/8 and 1/4 respectively, which colour of ball is most likely to be chosen.

   (b) Three factories X, Y, Z, whose outputs are in the ratios 2:3:5 respectively, all produce identical items. The proportion of defective items produced by factory X is 0.03, whilst the corresponding proportions for Y and Z are 0.04 and 0.05 respectively. If the outputs of the three factories are pooled, what is the probability that an item selected at random is defective?

6. (a) Four boxes A, B, C and D each contain a mixture of black and white balls. A contains 4 white and 5 black, B contains 3 white and 6 black, C contains 5 white and 4 black and D contains 2 white and 7 black. An unbiased die whose faces are marked A,B,C,D,D,D is rolled, and a ball is then chosen at random from the box which corresponds to the face showing on the die. Given that the chosen ball was white, find the conditional probability that the die showed B. What conclusion can we come to about the events "getting a B on the die" and "choosing a white ball"?

(b) In a survey, where the proportion of men to women questioned was 60:40, 60% of the men and 50% of the women said they smoked. If an individual chosen at random from this survey was found to be a non-smoker, find the probability that this person was a woman.

MODULE 5

Formulae and Worked Examples

5.1 Important results and formulae

In this section we have collected together the important definitions, results and formulae of Modules 1, 2, 3 and 4.

5.1.1. The number of **permutations**, or **ordered arrangements** of r objects taken from n unlike objects is written ^nP_r where

$$^nP_r = \frac{n!}{(n-r)!} \quad \text{(see 1.2.1)}$$

5.1.2 The number of **combinations**, or **unordered arrangements** of r objects taken from n unlike objects is written ^nC_r or $\binom{n}{r}$ where

$$^nC_r = \frac{n!}{r!(n-r)!} \quad \text{(see 1.3.5)}$$

5.1.3 Let A and B be two events. Then

$$P(A|B) = \frac{P(A \cap B)}{P(B)} \quad \text{(see 2.1.2)}$$

or equivalently

5.1.4.
$$P(A \cap B) = P(B) \times P(A|B) \quad \text{(see 2.2.1)}$$

Furthermore

5.1.5
$$P(A|B) = \frac{P(A) \times P(B|A)}{P(B)} \quad \text{(see 2.4.1)}$$

5.1.6. If C is a third event, then

$$P(A \cap B \cap C) = P(A) \times P(B|A) \times P(C|A \cap B) \quad \text{(see 2.2.6)}$$

5.1.7. In general, if $A_1, A_2, A_3, \ldots A_n$ are n events then

$$P(A_1 \cap A_2 \cap \ldots \cap A_n) =$$
$$P(A_1) \times P(A_2|A_1) \times P(A_3|A_1 \cap A_2) \times \ldots \times P(A_n|A_1 \cap A_2 \cap \ldots \cap A_{n-1})$$
$$\text{(see solution to 2.2.8.)}$$

5.1.8 Two events A and B are **independent** if any one (and consequently all three) of the following equations are satisfied:

$$P(A \cap B) = P(A).P(B) \quad \text{(see 3.2.2)}$$

$$P(A|B) = P(A) \quad \text{(see 3.2.1)}$$

$$P(B|A) = P(B) \quad \text{(see 3.2.3)}$$

5.1.9 Three events A, B and C are (totally) **independent** if and only if all the following conditions are satisfied.

$$P(A \cap B) = P(A).P(B)$$
$$P(A \cap C) = P(A).P(C)$$
$$P(B \cap C) = P(B).P(C)$$
$$P(A \cap B \cap C) = P(A).P(B).P(C) \quad \text{(see 3.4.1)}$$

5.1.10 In general, a **necessary** condition for n events $A_1, A_2, \ldots, A_n$ to be independent is that

$$P(A_1 \cap A_2 \cap \ldots \cap A_n) = P(A_1).P(A_2) \ldots P(A_n)$$

This is **not** a **sufficient** condition. (see 3.5.1)

Let $\{A_1, A_2, \ldots, A_n\}$ be a set of mutually exclusive and exhaustive events. Let A_r be any member of this set and let B be any other event. Then

5.1.11

$$P(B) = P(A_1) \times P(B|A_1) + P(A_2) \times P(B|A_2) + \ldots + P(A_n) \times P(B|A_n)$$
(The Law of Total Probability, see 4.2.1)

5.1.12

$$P(A_r | B) = \frac{P(A_r) \times P(B|A_r)}{P(A_1) \times P(B|A_1) + P(A_2) \times P(B|A_2) + \ldots + P(A_n) \times P(B|A_n)}$$

(Bayes' Theorem, see 4.3.2.)

5.2 Solutions to Examination Questions

The following worked examples have been selected form W.J.E.C, A3 Mathematics papers. No questions have been included from papers set since 1982 since these are readily available elsewhere.

Only one method of solution is given for each question, although in many cases alternative solutions may exist. Thus, in particular, some questions which have been solved by the use of combinations could equally well be tackeld by multiplying probabilities and vice versa. Similarly, Probability Tree Diagrams can often be used instead of Bayes' Theorem, and Venn Diagrams will usually simplify the solution of problems which would otherwise be solved by the algebraic manipulation of probabilities.

In general, we have attempted to exhibit as many varied methods of proof as possible.

1972 A3 Question 3

(a) Express $P(A \cup B)$ in terms of $P(A)$ and $P(B)$ in each of the cases when A and B are

  (i) mutually exclusive
  (ii) independent

(b) Suppose that a letter sent by first class mail has probability 0.6 of being delivered the next day, probability of 0.3 of being delivered the second day after the letter was posted, and probability 0.1 of being delivered on the third day after the letter was posted. Suppose further, that for a letter sent by second class mail, the corresponding probabilities are 0.2, 0.3 and 0.5 respectively.

Two letters are posted simultaneously, one by first class mail, and the other by second class mail. Assuming that the delivery days of the letters are independent, calculate the probabilities that

  (i) at least one of the two letters will be delivered the following day.
  (ii) the letter sent by first class mail will be delivered a day earlier than the letter sent by second class mail.

If three letters are posted simultaneously by second class mail find the probability that all three letters will be delivered on the same day.

Solution

(a) In general,
$P(A \cup B) = P(A) + (P(B) - P(A \cap B)$

(i) If A and B are mutually exclusive then

$P(A \cap B) = 0$

Thus $P(A \cup B) = P(A) + P(B)$

(ii) If A and B are independent, then

$P(A \cap B) = P(A).P(B)$
Thus $P(A \cup B) = P(A) + P(B) - P(A).P(B)$

(b) (i) P(at least one letter will be delivered the next day)
= 1 − P (neither letter will be delivered the next day)
= 1 − P (1st class letter will not be delivered the next day) x
P (2nd class letter will not be delivered the next day)
= 1 − 0.4 x 0.8
= 1 − 0.32
= 0.68

(ii) P (1st class letter delivered on 1st day and 2nd class letter delivered on 2nd day)

= P (1st class letter delivered on 1st day) x
P (2nd class letter delivered on 2nd day)
= 0.6 x 0.3 = 0.18
Similarly, P (1st class letter delivered on 2nd day and 2nd class letter delivered on 3rd day)
= 0.3 x 0.5 = 0.15

Therefore
P(1st class letter will be delivered a day earlier than 2nd class letter) = 0.18 + 0.15 = 0.33

P (all 3 letters are delivered on 1st day) = $(0.2)^3$ = 0.008
P (all 3 letters are delivered on 2nd day) = $(0.3)^3$ = 0.027
P (all 3 letters are delivered on 3rd day) = $(0.5)^3$ = 0.125
P (all 3 letters delivered on the same day)

= 0.008 + 0.027 + 0.125
= 0.16

1972 A3 Question 4

Three girls, two of whom are sisters, and five boys, two of whom are brothers, meet to play tennis. They draw lots to determine how they should split up into two groups of four to play doubles.

(a) Calculate the probabilities that one of the two groups will consist of

(i) boys only
(ii) two boys and two girls
(iii) the two brothers and the two sisters.

(b) If the lottery is organised so as to ensure that one of the two groups consists of two boys and two girls, calculate the probability that the two brothers and the two sisters will be in the same group. Given that the two brothers are in the same group, calculate the probability that the two sisters are also in that group.

Solution

(a) (i) P (1st group will be boys only) $= \dfrac{^5C_4}{^8C_4} = \dfrac{5}{70} = \dfrac{1}{14}$

P (one of the groups will be boys only) $= 2 \times 1/14 = 1/7$

(ii) Similarly
P (one of the groups will consist of 2 boys and 2 girls)
$= \dfrac{2 \times {}^5C_2 \times {}^3C_2}{{}^8C_4} = \dfrac{2 \times 10 \times 3}{70} = \dfrac{6}{7}$

(iii) P (one of the groups will contain the 2 brothers and 2 sisters)
$= 2 \times \dfrac{1}{{}^8C_4} = 2 \times \dfrac{1}{70} = \dfrac{1}{35}$

(b) The number of different ways in which a group can consist of 2 boys and 2 girls
$= {}^5C_2 \times {}^3C_2$
$= 10 \times 3$
$= 30$

P(the 2 brothers and 2 sisters will be in the same group)
$= 1/30$

The number of different ways in which a group can consist of the 2 brothers and 2 girls
$= {}^3C_2 = 3$

The number of different ways in which a group can consist of the 2 brothers, another boy and a girl (in this case the other group is the one consisting of 2 boys and 2 girls)
$= {}^3C_1 \times {}^3C_1 = 9$

P (the two sisters will be in the same group as the brothers | the brothers are in the same group)
$= \dfrac{1}{3 + 9} = \dfrac{1}{12}$

1973 A3 Question 3

Twelve athletes have entered for a race to be run on a six-lane track. Six of the athletes are drawn at random to run in the first heat, the remaining six to run in the second heat. If four of the athletes are Welshmen, calculate the probability that they will all run in the same heat.

The draw having been made, it transpires that all four Welshmen are in the same heat. Only the first three runners in a heat qualify for the final event. Assuming that the six runners in a heat have equal probabilities of qualifying for the final event, calculate the probabilities that exactly one, two and three of the Welshmen will qualify for the final event.

Assuming futher that, independently of the results in the heats, each Welshman who qualifies for the final event has probability 1/10 of winning the event, calculate the probability that the final event will be won by a Welshman.

Solution

P (all four Welshmen run in the first heat)

$= 4/12 \times 3/11 \times 2/10 \times 1/9 \times {}^6C_4 = 1/33$

P (all four Welshmen run in the same heat)

$= 1/33 + 1/33 = 2/33$

Now assume that all Welshmen are in the same heat.

Let W_r be the event that r Welshmen reach the final   r = 1,2,3

Then $P(W_1) = 4/6 \times 2/5 \times 1/4 \times {}^3C_1 = 1/5$

$P(W_2) = 4/6 \times 3/5 \times 2/4 \times {}^3C_1 = 3/5$

$P(W_3) = 4/6 \times 3/5 \times 2/4 = 1/5$

Let A be the event that the final is won by a Welshman
Then by the Law of Total Probability

$P(A) = P(A|W_1).P(W_1) + P(A|W_2).P(W_2) + P(A|W_3).P(W_3)$

$= 1/10 \times 1/5 \quad + \quad 2/10 \times 3/5 \quad + \quad 3/10 \times 1/5 = 1/5$

1973 A3 Question 4

A fair coin is tossed n ($\geq$ 2) times. Let A denote the event that there will be at least one head and at least one tail, and let B denote the event that there will be at most one head. Show that A and B are independent if n = 3, but are not independent for any other n $\geq$ 2.

Determine the minimum value of n if the probability of throwing at least one head and at least one tail in n tosses is to exceed 0.9.

Solution

P (no head amongst the n tosses) $= (1/2)^n$
P (no tails amongst the n tosses) $= (1/2)^n$
$P(A) = 1 - (1/2)^n - (1/2)^n = 1 - (1/2)^{n-1}$
P (exactly one head amongst the n tosses)
$= (1/2)^n \times {}^nC_1 = n \times (1/2)^n$
Therefore $P(B) = (1/2)^n + n(1/2)^n = (n+1)(1/2)^n$
But $A \cap B$ represents the event of getting exactly one head amongst the n tosses,
and thus
$P(A \cap B) = n \times (1/2)^n$
A and B will be independent if and only if
$\quad P(A \cap B) = P(A).P(B)$
if and only if
$\quad n \times (1/2)^n = \{1 - (1/2)^{n-1}\}\{(n+1)(1/2)^n\}$
if and only if
$\quad n = (n+1)(1 - (1/2)^{n-1})$
if and only if
$\quad n = n - n(1/2)^{n-1} + 1 - (1/2)^{n-1}$
if and only if
$\quad (n+1)(1/2)^{n-1} = 1$
if and only if
$\quad n + 1 = 2^{n-1}$

By inspection n = 3 is a root of this equation. Furthermore, by drawing the graphs of $f(x) = x + 1$ and $g(x) = 2^{x-1}$ on the same axes we see that the above equation has only one positive root.

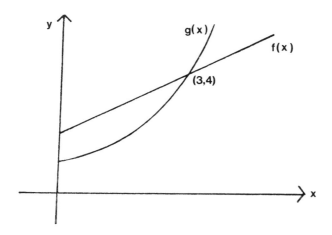

Therefore A and B are independent only when n = 3

We seek the minimum value of n such that P(A) > 0.9
    i.e.   such that $1 - (1/2)^{n-1} > 0.9$
    i.e.   such that $0.1 > (1/2)^{n-1}$
    i.e.   such that $2^{n-1} > 10$

Clearly, by inspection, n = 5 is the required value.

1974 A3 Question 2

Four cards are drawn at random, without replacement, from a pack of nine cards which are numbered from 1 to 9 respectively.

(a) Calculate the probability that

    (i)   both the numbers 1 and 9 will be drawn
    (ii)   the largest number drawn will be 8
    (iii) at least 3 even numbers will be drawn

(b) Given that the largest number drawn was 8, calculate the conditional probability that the smallest number drawn was 3.

(c) If the drawn cards are set down in a row from left to right as they are drawn, calculate the probability the resulting four digit number will be less than 5941.

Solution

(a)

(i) If both 1 and 9 are drawn, the other 2 cards can be any 2 from 7.

$$\therefore \text{ P (Both 1 and 9 are drawn)} = \frac{{}^7C_2}{{}^9C_4} = \frac{21}{126} = \frac{1}{6}$$

(ii) If the largest number drawn is 8, the other 3 cards can be any 3 from 7.

$$\therefore P\text{ (the largest number drawn is 8)} = \frac{^7C_3}{^9C_4} = \frac{35}{126} = \frac{5}{18}$$

(iii)
$$P\text{ (3 even numbers are drawn)} = \frac{^4C_3 \times {}^5C_1}{^9C_4} = \frac{20}{126}$$

$$P\text{ (4 even numbers are drawn)} = \frac{1}{^9C_4} = \frac{1}{126}$$

$\therefore$ P (at least 3 even numbers are drawn)

$= 20/126 + 1/126 \qquad = 21/126 \quad = \quad 1/6$

(b) We have already shown in (a) (ii) that there are $^7C_3 = 35$ ways in which the largest number drawn is 8.

If 3 is the smallest number drawn, the other 2 numbers can be any 2 from 4.

$\therefore$ P. (3 is the smallest number drawn|8 is the largest number)

$= \quad ^4C_2/35 \quad = \quad 6/35$

(c) Consider those possible 4 digit numbers which are less than 5000.
The first digit can be any one of four, while the final 3 may be any of $^8P_3$. Thus there are $4 \times 8 \times 7 \times 6 = 1344$ such possible numbers.

Consider those possible 4 digit numbers greater than 5000 but less than 5900. The first digit is fixed, the second digit can be any one of 7 and the final 2 digits can be any of $^7P_2$. Thus there are $7 \times 7 \times 6 = 294$ such possible numbers.

Consider those possible 4 digit numbers greater than 5900 but less than 5940. The first two digits are fixed, the third may be any one of three, and the last any one of 6. Thus there are $3 \times 6 = 18$ such possible numbers.

Thus the total number of possible 4 digit numbers less than 5941 $= 1344 + 294 + 18 = 1656$

The total number of possible 4 digit numbers $= 9 \times 8 \times 7 \times 6 = 3024$

Therefore

P (chosen number is less than 5941) $= 1656/3024 \quad = 23/42$

1974 A3 Question 3

Of the total daily production of a standard article at a certain factory, 40% of the articles are produced on machine M_1, 10% are produced on machine M_2, and 50% are produced on machine M_3. All articles produced on M_1 are coloured red. Of the articles produced on M_2, 30% are coloured red, 50% are coloured blue and 20% are coloured green, and of the articles produced on M_3, 50% are coloured red, 20% are coloured blue and 30% are coloured green.

(i) Calculate the probability that an article chosen at random from a day's production will not be red.
(ii) If three articles are drawn at random from articles produced on M_2, calculate the probability that they will consist of one blue and two green articles.
(iii) Three articles are drawn at random from articles produced on one of the machines and are found to consist of one blue and two green articles. Calculate the probability that they were produced on M_2.

Solution

(i) Using the law of Total Probability,
P (choosing a red article)
= 100/100 x 40/100 + 30/100 x 10/100 + 50/100 x 50/100
= 68/100   = 0.68
Therefore
P (not choosing a red article) = 0.32

(ii) P (one blue and two green articles | the machine was M_2)
50/100 x 20/100 x 20/100 x $^3C_1$ = 3/50 (= 30/500)

(iii) P (one blue and two green articles | the machine was M_1) = 0
P (one blue and two green articles | the machine was M_3)
= 20/100 x 30/100 x 30/100 x $^3C_1$ = 27/500

Hence by Bayes' Theorem,

P (the machine was M_2 | one blue and two green articles)

$$= \frac{10/100 \times 30/500}{0 + 10/100 \times 30/500 + 50/100 \times 27/500}$$

$$= \frac{300}{300 + 1350} = \frac{300}{1650}$$

= 2/11

1975 A3 Question 3

Three balls are drawn at random without replacement from a box containing 8 white, 4 black and 4 red balls. Calculate the probabilities that they will consist of

(i) at least one white ball,
(ii) two white balls and one black ball,
(iii) two balls of one colour, and another of a different colour,
(iv) one ball of each colour.

If instead, each ball is replaced in the box before the next ball is drawn, calculate the probability that the three balls drawn will consist of one of each colour.

Solution

(i) P(no white ball) = 8/16 x 7/15 x 6/14 = 1/10
Hence   P (at least one white ball) = 9/10

(ii) P (two white balls and one black ball) = 8/16 x 7/15 x 4/14 x $^3C_1$ = 1/5

(iii) As in (ii)
P (two white balls and one red ball) = 1/5
P (two red balls and one white ball)
= 4/16 x 3/15 x 8/14 x $^3C_1$ = 3/35

Similarly
P (two black balls and one white ball) = 3/35
P (two black balls and one red ball)
= 4/16 x 3/15 x 4/14 x $^3C_1$ = 3/70

Similarly

P (two red balls and one black ball) = 3/70

Therefore

P (two balls of one colour and one of a different colour)
= 1/5 + 1/5 + 3/35 + 3/35 + 3/70 + 3/70 = 46/70 = 23/35

(iv) P (one ball of each colour)
= 8/16 × 4/15 × 4/14 × 3!   = 8/35

If each ball is replaced before the next one is removed, then
P (one ball of each colour)

= 8/16 × 4/16 × 4/16 × 3!   = 3/16

1975 A3 Question 4 (parts (i) and (ii))

A census showed that 20% of all married couples living in a certain district had no child under 16 years of age, 50% had one child under 16, and 30% had 2 or more children under 16. The census also showed that both the husband and wife were in employment in 70% of couples having no child under 16, in 30% of couples having one child under 16 and in 10% of couples having more than one child under 16.

(i) If a married couple is chosen at random, calculate the probability that both husband and wife are in employment.
(ii) If a married couple is chosen at random from those couples where both the husband and wife are in employment, calculate the probability that the couple has at least one child under 16.

Solution

Let C_0 be the event that a married couple has no child under 16
  C_1 the event that a married couple has 1 child under 16
  C_2 the event that a married couple has 2 or more children under 16
and E the event that both husband and wife are in employment.

Then we are given
  $P(C_0)$ = 20/100 $P(E \mid C_0)$ = 70/100
  $P(C_1)$ = 50/100 $P(E \mid C_1)$ = 30/100
  $P(C_2)$ = 30/100 $P(E \mid C_2)$ = 10/100

(i) We require P(E)

By the Law of Total Probability

$P(E) = P(C_0) \cdot P(E \mid C_0) + P(C_1) \cdot P(E \mid C_1) + P(C_2) \cdot P(E \mid C_2)$

= 20/100 × 70/100 + 50/100 × 30/100 + 30/100 × 10/100

= 32/100   = 0.32

(ii) We require $P(C_0' \mid E)$

Now $P(C_0 \mid E) = \dfrac{P(C_0) \cdot P(E \mid C_0)}{P(E)}$

$= \dfrac{20/100 \times 70/100}{32/100} = \dfrac{14}{32}$

$P(C_0' \mid E) = 1 - 14/32 = 9/16$

1976 A3 Question 1

A box A contains 10 balls, 4 of which are white and 6 are black. If 3 balls are drawn at random without replacement from the box, calculate the probabilities that they will consist of

(i) three balls having the same colour,
(ii) more white than black balls.

Suppose that 2 balls are drawn at random from the original 10 balls in A and placed in a box B which originally contained 6 balls, 3 of which were white and 3 were black. If a ball drawn at random from the 8 balls which are in B is found to be black, calculate the probability that the 2 balls that were drawn from A and placed in B were both black.

Solution
(i)   P (choosing 3 white balls) = 4/10 x 3/9 x 2/8   = 1/30
      P (choosing 3 black balls) = 6/10 x 5/9 x 4/8   = 1/6
Thus
      P (choosing 3 balls of the same colour) = 1/30 + 1/6 = 1/5

(ii) P (choosing 2 white balls and 1 black ball)
      = 4/10 x 3/9 x 6/8 x $^3C_1$ = 3/10

So P (choosing more white balls than black balls)
      = 1/30 + 3/10   = 1/3

Let X_1 be the event of choosing 2 black balls from A
    X_2 be the event of choosing 1 white and 1 black ball from A
and X_3 be the event of choosing 2 white balls from A

Then $P(X_1)$ = 6/10 x 5/9 = 1/3 (= 5/15)
     $P(X_2)$ = 6/10 x 4/9 x $^2C_1$ = 8/15
     $P(X_3)$ = 4/10 x 3/9 = 2/15

Furthermore, X_1 will result in B having 5 black balls and 3 white balls. Thus if W denotes the event of removing a black ball from B, then
     $P(W|X_1)$ = 5/8

Similarly
     $P(W|X_2)$ = 4/8
and $P(W|X_3)$ = 3/8

Thus by Bayes' Theorem,

$$P(X_1|W) = \frac{P(X_1).P(W|X_1)}{P(X_1).P(W|X_1)+P(X_2).P(W|X_2)+P(X_3).P(W|X_3)}$$

$$= \frac{5/15 \times 5/8}{5/15 \times 5/8 + 8/15 \times 4/8 + 2/15 \times 3/8} = \frac{25}{25+32+6} = \frac{25}{63}$$

1976 A3 Question 2

(a) Two independent events are such that there is a probability of 1/6 that they will both occur and a probability of 1/3 that neither will occur. Calculate their individual probabilities of occuring.

(b) A fair cubical die has three of its faces coloured red, two coloured blue and one coloured white. If the die is thrown six times, calculate the probabilities that

   (i)   a red face will be uppermost at least once,

(ii) a red face will be uppermost exactly three times,
(iii) each colour will be uppermost exactly twice.

Solution

(a) Let the events be A and B
$P(A \cap B) = 1/6 \quad P(A \cup B) = 1 - 1/3 = 2/3$
However
$P(A) + P(B) = P(A \cup B) + P(A \cap B)$
and thus
$P(A) + P(B) = 2/3 + 1/6 = 5/6$ ..... (1)
Since A and B are independent
$P(A \cap B) = P(A) \cdot P(B)$
and thus $P(A) \cdot P(B) = 1/6$..... (2)

Substituting from 2 in 1, we have
$$P(A) + \frac{1}{6P(A)} = \frac{5}{6}$$

Therefore
$6P(A)^2 - 5P(A) + 1 = 0$
$(3P(A) - 1)(2P(A) - 1) = 0$

Hence
$P(A) = 1/3$ or $1/2$

If $P(A) = 1/3$, then $P(B) = 1/2$ and
if $P(A) = 1/2$, then $P(B) = 1/3$

∴ Required probabilities are 1/2 and 1/3

(b)
(i) P(no red faces) = $(3/6)^6$ = 1/64
Therefore
P(at least one red face) = $1 - 1/64 = 63/64$

(ii) P(exactly three red faces)

= P(three red faces and three faces of another colour)

= $(3/6)^3 \times (3/6)^3 \times {^6C_3}$

= 20/64 = 5/16

(iii) P(each colour will be uppermost twice)

= $(3/6)^2 \times (2/6)^2 \times (1/6)^2 \times {^6C_2} \times {^4C_2}$

= $1/4 \times 1/9 \times 1/36 \times 15 \times 6$

= 5/72

1977 A3 Question 1

(a) Two events A and B are such that $P(A) = 1/3$ and $P(B) = 1/2$. If A' denotes the complement of A, calculate $P(A' \cap B)$ in each of the cases when

(i) $P(A \cap B) = 1/8$
(ii) A and B are mutually exclusive
(iii) A is a subset of B

(b) A Scottish court may be given any one of the three verdicts 'guilty' 'not guilty' and 'not proven'. Of all the cases tried by the court, 70% of the verdicts are 'guilty', 20% are 'not guilty' and 10% are 'not proven'. Suppose that when the court's verdict is 'guilty', 'not guilty' and 'not proven', the probabilities that the accused is really innocent are 0.05, 0.95 and 0.25 respectively. Calculate the probability that an innocent person will be found 'guilty' by the court.

Solution

(a)
(i)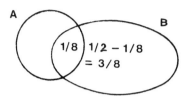

(ii)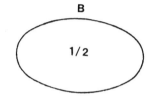

(iii)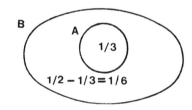

(b) Let I be the event that a person is innocent
X_1 the event that the court's verdict is 'guilty'
X_2 the event that the court's verdict is 'not guilty' and
X_3 the event that the court's verdict is 'not proven'

We must find $P(X_1 | I)$

We are given

$P(X_1) = 7/10 \quad P(I|X_1) = 5/100$
$P(X_2) = 2/10 \quad P(I|X_2) = 95/100$
$P(X_3) = 1/10 \quad P(I|X_3) = 25/100$

By Bayes' Theorem

$$P(X_1 | I) = \frac{P(X_1) \cdot P(I|X_1)}{P(X_1) \cdot P(I|X_1) + P(X_2) \cdot P(I|X_2) + P(X_3) \cdot P(I|X_3)}$$

$$= \frac{7/10 \times 5/100}{7/10 \times 5/100 + 2/10 \times 95/100 + 1/10 \times 25/100}$$

$$= \frac{35}{35 + 190 + 25} = \frac{35}{250} = \frac{7}{50}$$

1978 A3 Question 1

Each of two boxes contains ten discs. In one box, four of the discs are red, two are white and four are blue. In the other box two are red, three are white and five are blue. One of these two boxes is chosen at random, and three discs are drawn at random from it without replacement.

Calculate

(i) the probability that one disc of each colour will be drawn
(ii) the probability that no white disc will be drawn
(iii) the most probable number of white discs that will be drawn.

Given that three blue discs were drawn, calculate the conditional probability that they came from the box that contained four blue discs.

Solution

Let X_1 be the event of choosing the box with four blue discs
and X_2 the event of choosing the box with five blue discs

Then $P(X_1) = P(X_2) = 1/2$

(i) $P(\text{one disc of each colour} \mid X_1) = \dfrac{{}^4C_1 \times {}^2C_1 \times {}^4C_1}{{}^{10}C_3}$

$= (4 \times 2 \times 4)/120 = 32/120$

$P(\text{one disc of each colour} \mid X_2) = \dfrac{{}^2C_1 \times {}^3C_1 \times {}^5C_1}{{}^{10}C_3}$

$= (2 \times 3 \times 5)/120 = 30/120$

Therefore by the Law of Total Probability,

$P(\text{one disc of each colour}) = 1/2 \times 32/120 + 1/2 \times 30/120$

$= 31/120$

(ii) $P(\text{no white disc} \mid X_1) = \dfrac{{}^8C_3}{{}^{10}C_3} = \dfrac{56}{120}$

$P(\text{no white disc} \mid X_2) \dfrac{{}^7C_3}{{}^{10}C_3} = \dfrac{35}{120}$

Therefore by the Law of Total Probability

$P(\text{no white disc}) = 1/2 \times 56/120 + 1/2 \times 35/120 = 91/240$

(iii) Similarly

$P(\text{one white disc}) = \dfrac{1}{2} \times \dfrac{{}^2C_1 \times {}^8C_2}{{}^{10}C_3} + \dfrac{1}{2} \times \dfrac{{}^3C_1 \times {}^7C_2}{{}^{10}C_3}$

$$= \frac{1}{2} \times \frac{2 \times 28}{120} + \frac{1}{2} \times \frac{3 \times 21}{120}$$

$$= 119/240$$

$$P(\text{two white discs}) = \frac{1}{2} \times \frac{{}^2C_2 \times {}^8C_1}{{}^{10}C_3} + \frac{1}{2} \times \frac{{}^3C_2 \times {}^7C_1}{{}^{10}C_3}$$

$$= \frac{1}{2} \times \frac{8}{120} + \frac{1}{2} \times \frac{3 \times 7}{120}$$

$$= 29/240$$

P(three white discs) $= 1/2 \times 0 + 1/2 \times 1/{}^{10}C_3$

$= 1/240$

Therefore most probable number of white discs drawn $= 1$

$$P(3 \text{ blue discs} | X_1) = \frac{{}^4C_3}{{}^{10}C_3} = \frac{4}{120}$$

$$P(3 \text{ blue discs} | X_2) = \frac{{}^5C_3}{{}^{10}C_3} = \frac{10}{120}$$

By Bayes' Theorem

$$P(X_1 | 3 \text{ blue discs}) = \frac{P(X_1).P(3 \text{ blue discs} | X_1)}{P(X_1).P(3 \text{ blue discs} | X_1) + P(X_2).P(3 \text{ blue discs} | X_2)}$$

$$= \frac{1/2 \times 4/120}{1/2 \times 4/120 + 1/2 \times 10/120}$$

$$= \frac{4}{4 + 10} = \frac{2}{7}$$

1978 A3 Question 2

Write down an equation in terms of probabilities corresponding to the statements
(i)   The events A and B are independent
(ii) The events A and B are mutually exclusive

The events A, B and C are such that A and B are independent and A and C are mutually exclusive. Given that

$P(A) = 0.4 \quad P(B) = 0.2 \quad P(C) = 0.3$
$P(B \cap C) = 0.1$

Calculate $P(A \cup B)$, $P(C | B)$ and $P(B | A \cup C)$

Also calculate the probability that one and only one of the events B, C will occur.

Solution

(i)   P(A).P(B) = P(A ∩ B)

(ii) P(A ∩ B) = 0 or equivalently P(A ∪ B) = P(A) + P(B)

Consider the Venn diagram below

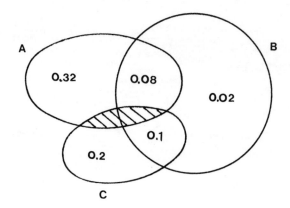

A ∩ C is shaded since A and C are mutually exclusive.
Since A and B are independent,

P(A ∩ B) = P(A).P(B) = 0.4 x 0.2 = 0.08

0.32 must be entered into the remaining part of A so that P(A) = 0.4

We are given that P(B ∩ C) = 0.1

The diagram may now be completed by using the fact that   P(B) = 0.2 and P(C) = 0.3

Thus:

P(A ∪ B) = 0.32 + 0.08 + 0.02 + 0.1 = 0.52

$$P(C|B) = \frac{0.1}{0.1 + 0.02 + 0.08} = 0.5$$

$$P(B|A \cup C) = \frac{0.08 + 0.1}{0.32 + 0.08 + 0.1 + 0.2} = \frac{0.18}{0.7} = \frac{9}{35}$$

P(one and only one of B, C)
= (0.08 + 0.02) + 0.2 = 0.3

1979 A3 Question 1

A pack of eight cards consists of the four aces and the four kings from a pack of ordinary playing cards.

(a) Two cards are dealt at random from this pack of cards.

(i)   Given that at least one of the cards drawn in an ace, calculate the probability that both cards are aces.

(ii) Given that one of the two cards drawn is the ace of spades, calculate the probability that the other card is an ace.

(b) Suppose now that the eight cards are shuffled and are then dealt one after the other.

   (i) Calculate the probability that the fifth card dealt will be the fourth ace dealt.
   (ii) Calculate the probability that the four aces will be dealt consecutively.

Solution

(a) No. of ways of dealing 2 aces = $^4C_2$ = 6
No. of ways of dealing no aces = $^4C_2$ = 6
Total number of ways dealing 2 cards = $^8C_2$ = 28
No. of ways of dealing at least one ace = 28 − 6 = 22

(i) P(two aces dealt | at least one is an ace) = 6/22
   = 3/11

(ii) If we know the ace of spades has been dealt, the other card can be any one of 7 cards, 3 of which are aces.

   P(two aces dealt | one is the ace of spades) = 3/7

(b)
   (i) P(5th card dealt will be the 4th ace dealt)
   = P(5th card is an ace and the 1st 4 cards contain 3 aces)
   = P(1st 4 cards contain 3 aces) x P(5th card is an ace | 1st 4 cards contain 3 aces.)
   = (4/8 x 3/7 x 2/6 x 4/5 x $^4C_3$) x 1/4
   = 8/35 x 1/4 = 2/35

   (ii) P(1st 4 cards will be aces) = 4/8 x 3/7 x 2/6 x 1/5 = 1/70
   Similarly, P(2nd, 3rd, 4th, 5th cards will be aces) = 1/70
   Therefore
   P(the four aces will be dealt consecutively)
   = 5 x 1/70 = 1/14

1979 A3 Question 2

Each of two bags A and B contain five white balls and four black balls, while a third bag C contains three white and six black balls.

(i) Suppose that one of the three bags was chosen at random and that two balls chosen at random without replacement from the chosen bag were both black. Calculate the probability that the chosen bag was C.

(ii) Suppose, instead that two of the three bags were chosen at random and that one ball was drawn at random from each of the chosen bags. Given that both balls drawn were black, calculate the probability that C was one of the chosen bags.

Solution

(i) Let X_1 be the event of choosing bag A
X_2 the event of choosing bag B
and X_3 the event of choosing bag C

Then $P(X_1) = P(X_2) = P(X_3) = 1/3$

Let Y be the event of drawing two black balls
Then
$$P(Y|X_1) = \frac{^4C_2}{^9C_2} = \frac{6}{36}$$

Similarly $P(Y|X_2) = 6/36$

$$P(Y|X_3) = \frac{^6C_2}{^9C_2} = \frac{15}{36}$$

By Bayes' Theorem

$$P(X_3 | Y) = \frac{P(X_3).P(Y|X_3)}{P(X_1).P(Y|X_1) + P(X_2).P(Y|X_2) + P(X_3).P(Y|X_3)}$$

$$= \frac{1/3 \times 15/36}{1/3 \times 6/36 + 1/3 \times 6/36 + 1/3 \times 15/36}$$

$$= \frac{15}{6 + 6 + 15} = \frac{5}{9}$$

(ii) Let Z_1 be the event of choosing A and B
Z_2 the event of choosing A and C
Z_3 the event of choosing B and C

Then $P(Z_1) = P(Z_2) = P(Z_3) = 1/3$

Let W be the event of drawing 2 black balls

$P(W|Z_1) = 4/9 \times 4/9 = 16/81$
$P(W|Z_2) = 4/9 \times 6/9 = 24/81$
$P(W|Z_3) = 4/9 \times 6/9 = 24/81$

P(C was one of the chosen bags | both balls are black)

$= P(Z_2|W) + P(Z_3|W)$

$$= \frac{P(Z_2).P(W|Z_2) + P(Z_3).P(W|Z_3)}{P(Z_1).P(W|Z_1) + P(Z_2).P(W|Z_2) + P(Z_3).P(W|Z_3)}$$

$$= \frac{1/3 \times 24/81 + 1/3 \times 24/81}{1/3 \times 16/81 + 1/3 \times 24/81 + 1/3 \times 24/81}$$

$$= \frac{24 + 24}{16 + 24 + 24} = \frac{3}{4}$$

1980 A3 Question 1

Four cards are to be drawn at random without replacement from a pack of ten cards numbered from 1 to 10 respectively.

(a) Calculate the probabilities that

(i) the largest number drawn will be 6
(ii) the product of the numbers will be even
(iii) all four numbers drawn will be consecutive integers.

(b) Given that at least two of the four numbers drawn were even, find the probability that every number drawn was even.

Solution

(a)
(i) The largest number chosen will be 6 if 6 is actually chosen and the other three numbers come from 1,2,3,4,5. This can be done in $^5C_3 = 10$ ways.

$$\therefore \text{P(largest number chosen is 6)} = \frac{10}{^{10}C_4} = \frac{10}{210} = \frac{1}{21}$$

(ii) The product of the numbers will be odd only if all four numbers are odd.

$$\therefore \text{P(the product is odd)} = \frac{^5C_4}{^{10}C_4} = \frac{5}{210} = \frac{1}{42}$$

$\therefore$ P(the product is even) = 41/42

(iii) P(numbers drawn are 1,2,3,4) = 1/210

Similarly, P(numbers drawn are 2,3,4,5) = 1/210, etc
P(numbers drawn are consecutive integers) = 7 x 1/210 = 1/30

(b) No. of ways of choosing 2 even and 2 odd numbers

$= {}^5C_2 \times {}^5C_2 = 10 \times 10 = 100$

No. of ways of choosing 3 even and 1 odd number

$= {}^5C_3 \times {}^5C_1 = 10 \times 5 = 50$

No. of ways of choosing 4 even numbers

$= {}^5C_4 = 5$

P(all numbers even | at least 2 are even)

$$= \frac{5}{100 + 50 + 5} = \frac{1}{31}$$

1980 A3 Question 2

(a) Two independent events A and B are such that P(A) = 0.4 and P(A∪B) = 0.7. Evaluate P(B) and P(A'∩B).

(b) In order to estimate what proportion of the pupils at a comprehensive school smoked, it was decided to interview a random sample of the pupils. To encourage truthful answers, each pupil in the sample was asked to toss a coin, and without divulging the outcome to the interviewer, to answer 'Yes' or 'No' to one of two questions dependent on the result of the toss. If the pupil had tossed a head, the question to be answered was 'Is your birthday in April?' while if the pupil has tossed a tail the question to be answered was 'Do you smoke?'. It is known that the proportion of pupils who were born in April is 0.1. Given that the proportion of pupils that answered 'Yes' was 0.2, estimate the proportion of pupils in the school who smoke.

Solution

(a) A and B are independent

Thus P(A∩B) = P(A).P(B) and hence
P(A) + P(B) = P(A).P(B) + P(A∪B)

Thus 0.4 + P(B) = 0.4 x P(B) + 0.7
P(B) (1 − 0.4) = 0.3
P(B) = 0.5

We can now display this information on the Venn Diagram below using initially the fact that
$P(A \cap B) = P(A).P(B) = 0.2$

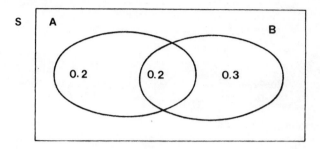

Thus from the diagram, $P(A' \cap B) = 0.3$

(b) Let A_1 be the event that the question asked was 'do you smoke?'
A_2 be the event that the question asked was 'were you born in April?'

Then $P(A_1) = P(A_2) = 1/2$
Let Y be event that the answer given was 'Yes'
We must calculate $P(Y \mid A_1)$ (assuming truthful answers)
We are given $P(Y) = 0.2$ and $P(Y \mid A_2) = 0.1$

Using the Law of Total Probability,
$P(Y) = P(Y \mid A_1).P(A_1) + P(Y \mid A_2).P(A_2)$
and thus
$0.2 = P(Y \mid A_1) \times 0.5 + 0.1 \times 0.5$

Thus $P(Y \mid A_1) = \dfrac{0.2 - 0.05}{0.5} = 0.3$

Assuming truthful answers, 0.3 would be an estimate for the proportion of pupils in the school who smoke.

1980 A3 Question 3 (part (a))

A factory has three machine A, B and C producing a certain type of item. Of the total daily output, 50% of the items are produced on A, 30% on B and 20% on C. The probabilities that an item produced on A, B and C are defective are 0.02, 0.02 and 0.07 respectively.

One item is chosen at random from a day's total output.

(i) Show that the probability that it will be defective is 0.03
(ii) Given that the chosen item is defective, calculate the probability that it was produced on machine C.

Solution

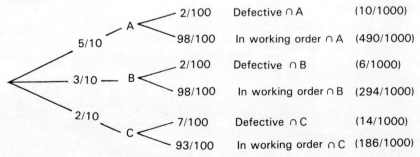

(i) P(choosing a defective item) = 10/1000 + 6/1000 + 14/1000 = 0.03
(ii) P(item was produced on C | item is defective)

$$\frac{14}{10+6+14} = \frac{7}{15}$$

1981 A3 Question 1

(a) The three events A, B and C have respective probabilities 2/5, 1/3 and 1/2. Given that A and B are mutually exclusive, P(A∩C) = 1/5 and P(B∩C) = 1/4.

(i) Show that only two of the three events are independent.
(ii) Evaluate P(C|B) and P(A'∩C')

(b) When Alec, Bert and Chris play a particular game, their respective probabilities of winning are 0.3, 0.1 and 0.6, independently for each game played. They agree to play a series of up to five games, the winner of the series (if any) to be the first player to win three games. Given that Bert wins the first two games of the series, show that

(i) Bert is just over 10 times more likely than Alec to win the series,
(ii) there is a slightly better than even chance that there will be a winner in the series.

Solution

(a) In the Venn Diagram below, A∩B is shaded since A and B are mutually exclusive. After then inserting the given values for P(A∩C) and P(B∩C), the remainder of the diagram may be completed.

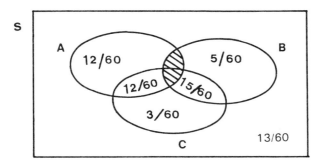

(i) P(A).P(C) = 2/5 x 1/2 = 1/5 = P(A∩C)
P(A).P(B) = 2/5 x 1/3 = 2/5 ≠ P(A∩B)
P(B).P(C) = 1/3 x 1/2 = 1/6 ≠ P(B∩C)

Thus the only two events which are independent are A and C

(ii) From the Venn diagram,

$$P(C|B) = \frac{15/60}{20/60} = \frac{3}{4}$$

P(A'∩C') = 5/60 + 13/60 = 18/60 = 3/10

(b) P(Bert wins the series in 3rd game) = 0.1
P(Bert wins the series in 4th game) = 0.9 x 0.1 = 0.09
P(Bert wins the series in 5th game) = 0.9 x 0.9 x 0.1 = 0.081
Therefore
P(Bert wins the series) = 0.271

P(Alec wins series) = P(Alec wins the 3 remaining games) = $(0.3)^3$ = 0.027

(i) P(Bert wins the series) is slightly greater than 10 x P(Alec wins the series)

P(Chris wins the series) = P(Chris wins the 3 remaining games) = $(0.6)^3$ = 0.216

(ii) P(there will be a winner to the series)
= 0.271 + 0.027 + 0.216
= 0.514

1981 A3 Question 2 (Part (a))

A batch of 20 items is inspected as follows. A random sample of 5 items is drawn from the batch without replacement and the number of defective items in the sample is counted. If this number is 2 or more the batch is rejected. If there is no defective item in the sample the batch is accepted. If there is exactly one defective item in the sample, then a further random sample of 5 items is drawn without replacement from the remaining 15 items in the batch. If this second sample includes at least 1 defective item then the batch is rejected, otherwise the batch is accepted.

Suppose that a batch to be inspected consists of exactly 2 defective items and 18 non-defective items.

Calculate the probabilities that

(i) the batch will be accepted on the basis of the first sample,
(ii) a second sample will be taken and the batch will then be accepted,
(iii) the batch will be rejected.

Solution

(i) P(batch will be accepted on the first sample)
= P(0 defective items among the 5)

$$\frac{^{18}C_5}{^{20}C_5} = \frac{18 \times 17 \times 16 \times 15 \times 14 / 120}{20 \times 19 \times 18 \times 17 \times 16 / 120} = \frac{21}{38}$$

P (batch will be rejected on the first sample)
= P (2 defective items among the 5)

$$\frac{^{18}C_3}{^{20}C_5} = \frac{18 \times 17 \times 16 / 6}{20 \times 19 \times 18 \times 17 \times 16 / 120} = \frac{1}{19}$$

(ii) P (a second sample will be taken)
= 1 − (21/38 + 1/19) = 15/38

P (batch is accepted | second sample is taken)

$$\frac{^{14}C_5}{^{15}C_5} = \frac{10}{15} = \frac{2}{3}$$

Therefore
P (a second sample is taken and the batch accepted)

= 15/38 x 2/3 = 5/19

(iii) P (batch is rejected | second sample is taken) = 1/3
P(a second sample is taken and batch rejected)

= 15/38 x 1/3 = 5/38

Hence
P (batch is rejected) = P (batch is rejected on 1st sample) + P (batch is rejected on 2nd sample)

= 1/19 + 5/38 = 7/38

1982 A3 Question 1

A box contains twelve balls numbered 1 to 12. The balls numbered 1 to 5 are red, those numbered 6 to 9 are white, and the remaining three balls are blue. Three balls are to be drawn at random without replacement from the box. Let A denote the event that each number drawn will be even, B the event that no blue ball will be drawn, and C the event that one ball of each colour will be drawn. Calculate (i) P(A) (ii) P(B) (iii) P(C) (iv) P(A∩C) (v) P(B∪C) (vi) P(A∪B).

Solution

(i) $P(A) = {}^6C_3 / {}^{12}C_3$ $= 20/220$ $= 1/11$

(ii) $P(B) = {}^9C_3 / {}^{12}C_3$ $= 84/220$ $= 21/55$

(iii) $P(C) = \dfrac{{}^5C_1 \times {}^4C_1 \times {}^3C_1}{{}^{12}C_3}$ $= \dfrac{60}{220}$ $= \dfrac{3}{11}$

(iv) $P(A \cap C) = (2 \times 2 \times 2)/220$ $= 8/220$ $= 2/55$

(v) $P(B \cap C) = 0$
     $P(B \cup C) = P(B) + P(C)$ $= 144/220$ $= 36/55$

(vi) $P(A \cap B) = {}^4C_3 / {}^{12}C_3$ $= 4/220$

Therefore
$P(A \cup B) = P(A) + P(B) - P(A \cap B)$
              $= 100/220$ $= 5/11$

1982 A3 Question 2 (1st part)

Ann and Bethan play a two-stage game. In the first stage two fair dice are thrown together. In the second stage, one of the girls draws a card at random from a pack of ten cards numbered 1 to 10. Ann draws the card at the second stage when the sum of the scores on the two dice is 6 or less, otherwise Bethan draws the card. Whoever draws the card wins the game if the number on the drawn card is 1 or 2, otherwise the other girl wins the game.

(i) Show that Ann's probability of winning the game is 0.55
(ii) Given that Ann wins the game, find the conditional probability that she drew the card.

Solution

Let X be the event that Ann draws the card at 2nd stage
    Y the event that Bethan draws the card at 2nd stage
    M the event that Ann wins the game
and N the event that Bethan wins the game

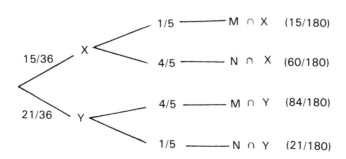

(i) P(Ann wins) = 15/180 + 84/180 = 99/180 = 0.55

(ii) P(Ann draws card | Ann wins) = $\dfrac{15}{15 + 84}$ = $\dfrac{5}{33}$

MULTIPLE CHOICE QUESTIONS

The following exercises consist of five different types of multiple choice questions. 3 questions of each type have been set for each module.

TYPE 1

A problem is set followed by 4 possible answers only one of which is correct. Write down the letter corresponding to the correct answer.

TYPE 2

A statement is made, possibly following on some given information. Write T if the statement is true or F if it is false.

TYPE 3

Some information is given, followed by 3 responses, one or more of which is necessarily implied by the information. Write down the letter(s) corresponding to all such responses.

TYPE 4

Each question contains two statements (a) and (b)
Write A if (a) implies (b) but (b) does not necessarily imply (a)
Write B if (b) implies (a) but (a) does not necessarily imply (b)
Write C if (a) implies (b) and (b) implies (a)
Write D if (a) denies (b) and (b) denies (a)
Write E if none of these relationships hold.

TYPE 5

Three bits of information are given in setting a problem. Do not solve the problem but decide whether:
(i) all the information is needed to solve the problem. In this case write A.
(ii) the information given is insufficient to solve the problem. In this case write I.
(iii) the problem can be solved without using one or more of the given pieces of information. In this case write down the letter(s) corresponding to the items of information not needed.

MODULE 1

TYPE 1

1. An even four digit number is to be constructed from the digits $\{2,3,4,5,6,8,\}$ in such a way that no digit may be used more than once. The number of ways in which this can be done is
   (a) 4^6 (b) $6!$ (c) $4 \times 5 \times 4 \times 3$ (d) $^6P_4$

2. 3 red balls, 2 white balls, and 4 green balls are placed in a bag. 3 balls are removed at random from the bag and placed in a row from left to right. The probability that they will be green, green, red in that order is
   (a) $\dfrac{^4C_2 \times {}^3C_1}{^9C_3}$ (b) $\dfrac{^4P_2 \times {}^3P_1}{^9P_3}$ (c) $\dfrac{1}{^9P_3}$ (d) $\dfrac{1}{^9C_3}$

3. A random sample of 3 pupils is taken from a class containing 10 boys and 9 girls. The most likely number of girls in the sample is
   (a) 0 (b) 1 (c) 2 (d) 3

TYPE 2

1. The number of different arrangements of the letters of the word PONTARDAWE is 10!

2. 2 red counters and 6 blue counters are placed in a box. 2 counters are removed at random from the box without replacement. The probability that both are red is 1/4.

3. Three different numbers are chosen at random from the set $\{1,2,3,4,5\}$. Their sum is more likely to be odd than even.

TYPE 3

1. A bag contains 7 red balls and 4 white balls. 2 balls are to be removed at random without replacement.
   (a) The probability that they will be of different colours is 28/55.
   (b) The probability that both will be red is 21/55.
   (c) The probability that both will be white is 7/55.

2. 4 different digits are chosen at random from the set $\{1,2,3,4,5,6,7,8\}$ and then placed in a row.
   (a) The probability that the 3rd digit is the 1 is 1/8.
   (b) The number of such arrangements is $^8C_4$.
   (c) The probability that the product of the four numbers is odd is 1/2.

3. A group of 10 people consists of 5 married couples. 2 are chosen at random from this group without replacement.
   (a) The probability that the two people are husband and wife is 1/9.
   (b) The probability that the two people are both men is 2/9.
   (c) The probability that the two people are male and female, but not husband and wife is 4/9.

TYPE 4

1. (a) $^nP_r = 210$
   (b) $^nC_r = 50$

2. A race consists of 8 runners, all of whom are either Welshmen or Englishmen.
   (a) The probability of at least one Welshman finishing in the first four is 1.
   (b) There are 5 Welshmen and 3 Englishmen.

3. A bag contains 12 counters all of which are either red or white. 3 counters are drawn at random from the bag without replacement.
   (a) The probability that the first counter is red is 2/3.
   (b) The probability that all 3 are red is 14/55.

TYPE 5

1. 20 cards are removed from an ordinary pack of playing cards. 6 cards are then removed at random without replacement from these 20 cards. Find the probability that 2 of the 6 cards are spades and another 2 are diamonds.
   (a) The original 20 cards contained 6 spades.
   (b) The original 20 cards contained 5 hearts.
   (c) The original 20 cards contained 3 diamonds.

2. A subcommittee is to be formed from the members of a full committe. Find the probability that the subcommittee will contain precisely 3 women.
   (a) The full committee has 15 members.
   (b) There are 9 women on the full committee.
   (c) The subcommittee has 5 members.

3. Find the probability that a Welshman will win a prize in next week's international cross-country event.
   (a) 4 Welshmen have qualified for the race.
   (b) All runners competing have an equal chance of winning.
   (c) Only the first three finishers win a prize.

MODULE 2

TYPE 1

1. A bag contains 7 white beads and 5 blue beads. 3 beads are removed at random without replacement. Given that at least one of the beads removed is white, the conditional probability that all three are white is (a) $^7C_3/^{12}C_3$ (b) $^7C_3/(^5C_2 \times ^7C_1)$ (c) $1/^7C_3$ (d) $^7C_3/(^{12}C_3 - ^5C_3)$

2. 30% of the pupils in a particular fifth form take History, and 2/3 of this History class speak Welsh. The probability that one of these Welsh speaking History students will sit his/her examination in Welsh is 0.75. If a pupil is chosen at random from this fifth form, the probability that he/she will be a Welsh speaking History student who will sit his/her examination in Welsh is (a) 0.75 (b) 0.5 (c) 0.15 (d) 0.3

3. The probability that I will score in any game for my village football team is 1/4, and we seem to win 1/3 of those games in which I score. Overall, we win 1/2 of our games. Given that we won yesterday, the probability that I also scored is (a) 1/6 (b) 1/4 (c) 1/2 (d) 1/12.

TYPE 2

1. A and B are two events such that $P(A) = P(B)$
   Then $P(B|A) = P(A|B)$

2. Two children are chosen at random without replacement from a group of 7 boys and 8 girls. The probability that the first child chosen will be a boy and the second a girl is
   $$(^7C_1 \times ^8C_1)/^{15}C_2$$

3. A pack of cards is numbered 1−17. Three cards are removed at random without replacement. The probability that their product is odd is 21/170.

TYPE 3

1. A bag contains 6 red beads, 4 yellow beads and 2 green beads. Three beads are removed from the bag at random without replacement. The probability that the third bead removed will be red
   (a) Equals 1/2.
   (b) Cannot be calculated until the colours of the first two beads are known.
   (c) Is three times the probability that it will be green.

2. A and B are two events such that $P(A|B) = 1$.
   (a) $P(A) = 1$
   (b) $P(B) \leq P(A)$
   (c) $P(A \cap B) = P(A)$

3. All the diamonds are removed from a pack of ordinary playing cards, and three cards are then removed from the remaining 39. The probability that all three will be clubs is
   (a) $^{13}C_3/^{39}C_3$ (b) $13/39 \times 12/38 \times 11/37$ (c) $^{13}P_3/^{39}P_3$

TYPE 4

1. A, B and C are three events
   (a) $P(A \cap B \cap C) = 1/4$ (b) $P(A \cap B) = 1/3$

2. The probability that it will rain on any given day is 1/2.
   (a) When it rains, the probability that I will have taken my umbrella with me is 2/3.
   (b) When I take my umbrella with me, the probability that it will rain is 1/4.

3. The probability that I will have chips for lunch is 4/15.
   (a) The probability that I will have chips for lunch and ice-cream for pudding is 1/9.
   (b) On those days when I have chips for lunch, the probability that I will have ice cream for pudding is 5/12.

TYPE 5

1. A, B and C are three events. Find $P(A \cap B \cap C)$.
   (a) $P(A) = 1/2$ (b) $P(B|A) = 2/3$ (c) $P(C|B) = 1/5$

2. A child is chosen at random from among the pupils of a particular school and is found to have blond hair. Find the probability that he/she has blue eyes.
   (a) P(a pupil has blond hair and blue eyes) = 1/12.
   (b) P(a pupil has blue eyes) = 2/9.
   (c) P(a pupil has blond hair) = 1/6.

3. A and B are two events. Find $P(A \cup B)$.
   (a) $P(A) = 3/5$ (b) $P(B) = 3/10$ (c) $P(A|B) = 2/3$

MODULE 3

TYPE 1

1. A bag contains three red beads, 4 blue beads and 6 yellow beads. Three beads are removed at random without replacement. The probability that all three will be of different colours is:
   (a) 3/13 x 4/13 x 6/13
   (b) 3/13 x 4/12 x 6/11
   (c) 3! x 3/13 x 4/13 x 6/13
   (d) 3! x 3/13 x 4/12 x 6/11

2. A, B and C take turns at throwing dice, the first to throw a six being the winner. If A throws first, the probability that he will win on his second throw is
   (a) 1/6   (b) 1/6 x (5/6)^3 (c) 1/6 x 5/6   (d) 1/3

3. A rifleman shoots at four targets. if his probability of hitting any target is always 2/3, the probability that he will hit precisely 2 of the targets is
   (a) 1/2   (b) $(2/3)^2$ (c) $(2/3)^2$ x $(1/3)^2$ (d) $^4C_2$ x $(2/3)^2$ x $(1/3)^2$

TYPE 2

1. A and B are independent events. Then it follows necessarily that P(A|B) = P(B|A).

2. The probability of getting at least one six in four throws of an unbiased die is 671/1296.

3. An unbiased die is thrown once. Throwing an even number and throwing a number divisible by 3 are independent events.

TYPE 3

1. Two events A and B are independent.
   (a) P(A|B) = P(A)
   (b) P(B|A) = P(B)
   (c) P(A∩B) = P(A) + P(B)

2. A computer randomly generates whole numbers from 1 to 30 inclusive. If 6 such numbers are generated then,
   (a) The probability that both the first two numbers will be divisible by 5 is 1/25.
   (b) The probability that each of the first three numbers will be divisible by 4 is 1/64.
   (c) The probability that precisely 5 of the 6 numbers will be divisible by 3 is $(1/3)^5$ x (2/3).

3. A fair coin is tossed twice. Let A be the event that the first toss yields a head, B the event that the second toss yields a tail and C the event that exactly one toss yields a head.
   (a) A and B are independent events
   (b) A and C are independent events
   (c) A, B and C are totally independent events.

TYPE 4

1. A and B are independent events such that P(A) ≠ 0, P(B) ≠ 0.
   (a) A and B are independent
   (b) A and B are mutually exclusive.

2. A, B and C are three events.
   (a) A, B and C are totally independent
   (b) P(A∩B∩C) = P(A) x P(B) x P(C).

3. The probability that it will rain tomorrow is 1/3. The probability that it will be windy tomorrow is 1/4.
   (a) The probability that it will be both wet and windy tomorrow is 1/12.
   (b) If it is windy tomorrow, the probability that it will be dry is 2/3.

TYPE 5

1. Are the events A and B independent?
   (a) P(A) = 2/3
   (b) P(B) = 1/2
   (c) P(A∪B) = 1.

2. Find the probability that tomorrow morning I will get up late, miss my breakfast and be late for work. On any given day
   (a) The probability that I will get up late = 1/10
   (b) The probability that I will miss breakfast = 1/9
   (c) The probability that I will be late for work = 1/12.

3. A bag contains a mixture of red, white and yellow balls. Three balls are removed from the bag, each ball being replaced before the next is removed. Find the probability that all three balls are red.
   (a) The bag contains 20 balls
   (b) 1/4 of the balls are red
   (c) 8 of the balls are white.

MODULE 4

TYPE 1

1. The probability that both Dafydd and Mair will pass their mathematics examination is 3/5. The probability that Dafydd will pass and Mair will fail is 1/5. The probability that Dafydd will pass is
   (a) 4/5 (b) 2/5 (c) 3/5 (d) 3/25.

2. Students in a college language faculty study one of French, German or Italian. The probability that a student will pass in French is 5/8, while the corresponding probabilities for German and Italian are 1/2 and 3/4. If the ratios of the students studying these languages is 3:2:1, the probability that a language student chosen at random will pass his/her examination is
   (a) 5/8 (b) 15/64 (c) 29/48 (d) 1/3.

3. On Saturdays and Sundays, the probability that I will wake up before 8 a.m. is 1/4. On any other day of the week, the corresponding probability is 3/4, except Thursdays, when it is 1/2. On waking today, I looked at the clock and saw that it was 7.45 a.m. The conditional probability that it is a Thursday is
   (a) 1/2 (b) 1/4 (c) 1/8 (d) 1/7.

TYPE 2

1. A, B and C are exhaustive events, and D any other event. Then it follows necessarily that $P(D) = P(D \cap A) + P(D \cap B) + P(D \cap C)$.

2. The probability that it will be both wet and cold tomorrow is 1/6. The probability that it will be dry and cold tomorrow is 1/4. If it is cold tomorrow, then the probability that it will also be dry is 1/2.

3. A and B are two events
   then $P(B|A) = \dfrac{P(B).P(A|B)}{P(A \cap B) + P(A \cap B')}$

TYPE 3

1. $\{A_1, A_2, \ldots, A_n\}$ form a set of mutually exclusive and exhaustive events. If B is any event, then
   (a) $P(B) = P(B \cap A_1) + P(B \cap A_2) + \ldots + P(B \cap A_n)$
   (b) $P(B) = P(B|A_1) + P(B|A_2) + \ldots + P(B|A_n)$
   (c) $P(A_1|B) = \dfrac{P(A_1) \times P(B|A_1)}{P(B)}$

2. A, B and C are three events such that
   $P(A \cap B) = 0$, $P(A) = 2/3$, $P(B) = 1/3$, $P(C|A) = 1/4$, $P(C|B) = 3/4$
   Then
   (a) $P(C) = 5/12$
   (b) $P(A|C) = 2/5$
   (c) $P(B|C) = 3/5$

3. $\{A_1, A_2, A_3\}$ form a set of mutually exclusive and exhaustive events. B is any event.
   (a) $P(A_1) + P(A_2) + P(A_3) = 1$
   (b) $\{(B \cap A_1), (B \cap A_2), (B \cap A_3)\}$ is a set of mutually exclusive events
   (c) $\{(B \cap A_1), (B \cap A_2), (B \cap A_3)\}$ is a set of exhaustive events.

TYPE 4

1. A, B and C are three events such that $P(A \cap B) = 0$
   (a) $P(A|C) = \dfrac{P(A) \times P(C|A)}{P(A) \times P(C|A) + P(B) \times P(C|B)}$
   (b) $B = A'$

2. (a) Both men and women are equally divided in favour of and against a particular proposal.
   (b) The probability that a randomly chosen person will be in favour of the proposal is 1/2.

3. A, B, C and D are 4 events.
   (a) {A,B,C} are mutually exclusive.
   (b) $P(A \cap D) = 2/7$ (b) $P(B \cap D) = 1/3$ (c) $P(C \cap D) = 2/5$

TYPE 5

1. If I walk to school tomorrow, find the probability that I will be late.
   (a) I either walk to school or go by bus.
   (b) The probability that I will walk to school on any given day is 1/3.
   (c) On those days when I am late, the probability that I will have walked is 1/2.

2. Find the probability that a person chosen at random will vote Conservative.
   (a) 42% of women intend voting Conservative.
   (b) 38% of men intend voting Conservative.
   (c) The ratio of men to women in the population is 49/51.

3. B and C are two events. Find $P(C|B)$.
   (a) $P(C \cap B) = 1/2$
   (b) $P(C' \cap B) = 1/6$
   (c) $P(B|C) = 5/6$

Answers to self-assessment questions (Module 1)

1.1.5 (i) $4^3 = 64$ (ii) $4 \times 3 \times 2 = 24$

1.1.6 $3 \times 2 \times 1 = 6$

1.1.7 (i) $2 \times 5 \times 5 \times 5 \times 3 = 750$ (ii) $3 \times 3 \times 2 \times 1 + 2 \times 3 \times 2 \times 1 = 30$

1.1.8 (i) $5 \times 4 \times 3 \times 2 = 120$ (ii) $5 \times 5 \times 5 \times 5 = 625$

1.1.9 $2 \times 2 \times 2 \times 2 \times 2 = 32$

1.1.10 $2 \times 4 \times 3 \times 2 \times 1 = 48$

1.1.11 $10 \times 9 \times 8 \times 7 \times 6 \times 5 = 151200$

1.1.18 $15 \times 14 \times 12 \times 11 = 27720$

1.1.19 (i) $4 \times 3 \times 10 \times 9 \times 8 = 8640$
(ii) $4 \times 3 \times 10 \times 10 \times 10 = 12,000$
(iii) $4 \times 4 \times 10 \times 10 \times 10 = 16,000$

1.1.20 $8!/2! = 20,160$ (ii) $8!/(2! \times 2!) = 10,080$ (iii) $6!/(3! \times 2!) = 60$

1.1.21 (i) $(7!/2!)/10,080 = 2,520/10,080 = 1/4$
(ii) $7 \times 6 \times 6!/(2! \times 2!)/10,080 = 3/4$

1.1.22 $3 \times 4/60 = 1/5$

1.1.23 (i) $12!/(3! \times 4! \times 5!) = 27,720$
(ii) $9!/(2! \times 3! \times 4!)/27,720 = 1260/27,720 = 1/22$

1.1.24 (i) $7! = 5040$ (ii) $2 \times 6!/7! = 2/7$

1.1.25 (i) $7 \times 6/2 + 7 = 28$ (ii) $7/28 = 1/4$ (iii) $12/28 = 3/7$

1.1.26 (i) $7!/2 = 2520$ (ii) $(6 \times 5!)/2 / 2520 = 1/7$

1.2.4 $1/120$

1.2.5 $(^7P_5 \times {}^5P_1)/{}^{12}P_3 = (7 \times 6 \times 5)/(12 \times 11 \times 10) = 7/44$

1.2.6 $(^6P_2 \times {}^3P_1)/{}^{14}P_3 = (6 \times 5 \times 3)/(14 \times 13 \times 12) = 15/364$

1.2.7 (i) $(^4P_2 \times {}^4P_3)/{}^8P_5 = (4 \times 3 \times 4 \times 3 \times 2)/(8 \times 7 \times 6 \times 5 \times 4) = 3/70$
(ii) $(^6P_3 \times 5)/{}^8P_5 = (6 \times 5 \times 4 \times 5)/(8 \times 7 \times 6 \times 5 \times 4) = 5/56$

1.2.8 ${}^9P_3/{}^{11}P_5 = (9 \times 8 \times 7)/(11 \times 10 \times 9 \times 8 \times 7) = 1/110$

1.2.9 (i) $1/{}^8P_2 = 1/56$ (ii) $4 \times 1/56 = 1/14$ (iii) $2 \times 1/14 = 1/7$

1.3.10 ${}^9C_3 = 84$

1.3.11 ${}^{12}C_8 = 495$. This is a combination not a permutation.

1.3.12 $(1/2) \times {}^{10}C_5 = 126$

1.3.13 (i) ${}^7C_3 = 35$ (ii) ${}^4C_3/{}^7C_3 = 4/35$

1.3.14 $\dfrac{{}^3C_2}{{}^7C_2} = \dfrac{3}{21} = \dfrac{1}{7}$

1.3.15 (i) ${}^9C_4 = 126$

(ii) $\dfrac{{}^7C_2}{{}^9C_4} = \dfrac{21}{126} = \dfrac{1}{6}$

1.3.16 $\dfrac{{}^5C_2 \times {}^7C_4}{{}^{12}C_6} = \dfrac{25}{66}$

1.3.17 $\dfrac{{}^7C_3 \times {}^6C_2 + {}^7C_4 \times {}^6C_1 + {}^7C_5}{{}^{13}C_5} = \dfrac{756}{1287} = \dfrac{84}{143}$

1.3.18 $\dfrac{{}^4C_1 \times {}^3C_2 + {}^4C_3}{{}^7C_3} = \dfrac{4 \times 3 + 4}{35} = \dfrac{16}{35}$

1.3.19 (i) $\dfrac{{}^4C_3}{{}^9C_3} = \dfrac{4}{84} = \dfrac{1}{21}$

(ii) $\dfrac{3 \times 4 \times 2}{{}^9C_3} = \dfrac{24}{84} = \dfrac{2}{7}$

1.3.20 (i) $\dfrac{{}^{13}C_4 \times {}^{13}C_2}{{}^{52}C_6} = \dfrac{715 \times 78}{20358520} = 0.00274$

(ii) $\dfrac{{}^{13}C_2 \times {}^{13}C_2 \times {}^{13}C_2}{{}^{52}C_6} = \dfrac{78 \times 78 \times 78}{20358520} = 0.0233$

(iii) $\dfrac{{}^{13}C_3 \times {}^{39}C_3}{{}^{52}C_6} = \dfrac{286 \times 9139}{20358520} = 0.1284$

Answers to self-assessment questions (Module 2)

2.1.8. (i) Venn diagram: E ∩ M with E only = 3/12, intersection = 5/12, M only = 4/12

(ii) $\dfrac{5/52}{9/12} = 5/9$

2.1.9 $\dfrac{1/30}{2/5} = 1/12$

2.1.10 $\dfrac{10/150}{50/150} = \dfrac{10}{50} = \dfrac{1}{5}$

2.1.11 (i) 12/28 (ii) $\dfrac{7/28}{16/28} = 7/16$

2.1.12 $\dfrac{2/36}{1/2} = 1/9$ (ii) $\dfrac{2/36}{5/36} = 2/5$

2.1.13 $\dfrac{21/45}{42/45} = 21/42 = 1/2$

2.1.14 (i) $\dfrac{16/28}{22/28} = 16/22 = 8/11$ (ii) $\dfrac{6/22}{12/22} = 6/12 = 1/2$

2.1.15 $\dfrac{40/91}{11/13} = 40/77$

2.2.8   (ii) $P(A_1 \cap A_2 \cap A_3 \cap \ldots \cap A_n) = P(A_1) \times P(A_2|A_1) \times P(A_3|A_1 \cap A_2) \times \ldots \times P(A_n|A_1 \cap A_2 \cap \ldots \cap A_{n-1})$

2.2.12 $1/2 \times 2/3 = 1/3$

2.2.13 $16/28 \times 12/27 = 16/63$

2.2.14 $1/4 \times 2/5 = 1/10$

2.2.15 $3/5 \times 3/5 + 2/5 \times 1/3 = 37/75$

2.2.16 $2/12 \times 5/11 + 3/12 \times 6/11 = 28/132 = 7/33$

2.2.17   (i) $3/10 \times 5/9 \times 2/8 = 1/24$ (ii) $2/10 \times 3/9 \times 2/8 = 1/60$
         (iii) $5/10 \times 3/9 \times 2/8 \times {}^3C_1 = 1/8$

2.2.18 $2/7 \times 1/6 \times 6 = 2/7$

2.2.19 $9/16 \times 8/15 \times 7/14 \times 6/13 \times 7/12 \times 6/11 \times 5/10 \times {}^7C_4 = 441/1144$

2.3.3   (i) $1/4 \times 1/3 = 1/12$ (ii) $2 \times 1/4 \times 1/3 = 1/6$ (iii) $1/4 \times 2/3 = 1/6$

2.3.4 $1/5 \times 1/4 \times 1/3 \times 1/2 \times 1 = 1/120$

2.3.5 $7/12 \times 6/11 \times 5/10 = 7/44$

2.3.6 $6/14 \times 5/13 \times 3/12 = 15/364$

2.3.7   (i) $4/8 \times 3/7 \times 4/6 \times 3/5 \times 2/4 = 3/70$ (ii) $1/8 \times 1/7 \times 5 = 5/56$

2.3.8 $1/11 \times 1/10 = 1/110$

2.3.9   (i) $1/8 \times 1/7 = 1/56$ (ii) $4 \times 1/8 \times 1/7 = 1/14$ (iii) $8 \times 1/8 \times 1/7 = 1/7$

2.3.10 $6/11 \times 5/10 \times 4/9 \times 5/8 \times 4/7 \times 5!/(3! \times 2!) = 100/231$

2.3.11   (i) $5/9 \times 4/8 \times 3/7 \times 2/6 = 5/126$
         (ii) $5/9 \times 4/8 \times 3/7 \times 4/6 \times 4 = 20/63$
         (iii) $5/126 + 40/126 = 45/126 = 5/14$
         (iv) $1 - 5/126 = 121/126$

2.3.12 $4/7 \times 3/6 \times 2/5 = 4/35$

2.3.13 $3/7 \times 2/6 = 1/7$

2.3.14 $2/9 \times 1/8 \times 4!/(2! \times 2!) = 1/6$

2.3.15 $7/12 \times 6/11 \times 5/10 \times 4/9 \times 5/8 \times 4/7 \times 6!/(4! \times 2!) = 25/66$

2.3.16 $7/13 \times 6/12 \times 5/11 \times 6/10 \times 5/9 \times 5!/(3! \times 2!)$
         $+ \; 7/13 \times 6/12 \times 5/11 \times 4/10 \times 6/9 \times 5$
         $+ \; 7/13 \times 6/12 \times 5/11 \times 4/10 \times 3/9$
         $= 175/429 + 70/429 = 252/429 = 84/143$

2.3.17 $4/7 \times 3/6 \times 2/5 + 4/7 \times 3/6 \times 2/5 \times 3 = 4/35 + 12/35 = 16/35$

2.3.18   (i) $4/9 \times 3/8 \times 2/7 = 1/21$ (ii) $4/9 \times 3/8 \times 2/7 \times 6 = 2/7$

2.3.19   (i) $13/52 \times 12/51 \times 11/50 \times 10/49 \times 13/48 \times 12/47 \times 6!/(4! \times 2!) = 0.00274$
         (ii) $13/52 \times 12/51 \times 13/50 \times 12/49 \times 13/48 \times 12/47 \times 6!/(2! \times 2! \times 2!) = 0.0233$
         (iii) $13/52 \times 12/51 \times 11/50 \times 39/49 \times 38/48 \times 37/47 \times 6!/(3! \times 3!) = 0.1284$

2.4.3 $(2/5) \times (1/8) / (3/10) = 1/6$

2.4.4 $(3/4) \times (2/3) / (3/5) = 5/6$

2.4.5 $(2/5) \times (1/2) / (2/3) = 3/10$

2.4.6   (i) 0   (ii) 0   A and B are two mutually exclusive events

---

Answers to self-assessment questions (Module 3)

3.1.3   Independent

3.1.4   (a) Independent   (b) Not independent

3.1.5   Independent

3.1.6   Not independent

3.1.7   Independent

3.1.8   Not independent

3.1.9   Independent

3.2.5 $P(A) = 1/4$ $P(B) = 1/13$ $P(C) = 1/2$
        $P(A \cap B) = 1/52$ $P(B \cap C) = 1/26$ $P(A \cap C) = 0$

(i) $P(A).P(B) = P(A \cap B)$ A and B are independent
(ii) $P(B).P(C) = P(B \cap C)$ B and C are independent
(iii) $P(A).P(C) \neq P(A \cap C)$ A and C are not independent

3.2.6 $P(A) = 1/6$ $P(A|B) = 1/5$
        A and B are not independent

3.2.7 $P(M) = 1/5$, $P(S) = 1/3$, $P(S \cup M) = 1 - (8/15) = 7/15$
        hence $P(S \cap M) = 1/5 + 1/3 - 7/15 = 1/15$
        Since $P(S \cap M) = P(S).P(M)$, S and M are independent

3.2.8 $P(A \cap B) = 1/4 + 1/2 - 2/3 = 1/12$
        $P(A).P(B) = 1/4 \times 1/2 = 1/8$
        A and B are not independent

3.2.9 $P(R) = 1/3$, $P(W|R) = 1/4$, $P(W \cup R) = 1 - 1/2 = 1/2$
        $P(R \cap W) = 1/3 \times 1/4 = 1/12$
        $P(W) = 1/2 + 1/12 - 1/3 = 1/4$
        Since $P(W) = P(W|R)$ W and R are independent
        Therefore a high wind does not affect the probability that it will rain.

3.3.3 $1/2 \times 1/3 = 1/6$

3.3.4   (i) $1/10 \times 1/10 = 1/100$ (ii) $4/10 \times 3/10 = 3/25$
        (iii) $1 - 4/5 \times 4/5 = 9/25$

3.3.5   (i) $1/25 \times 1/25 = 1/625$ (ii) $24/25 \times 24/25 = 576/625$
        (iii) $1/25 \times 24/25 + 24/25 \times 1/25 = 48/625$

3.3.6 $99/100 \times 95/100 = 9405/10000$

3.3.7 $P(A).P(G) = 1/8$, $P(A) + P(G) - 1/8 = 19/24 \Rightarrow$ $P(A) = 3/4$

3.3.8 $P(C \cap I) = 1/5 \times 1/6 = 1/30$
        P(chips only) $= 1/5 - 1/30 = 5/30$
        P(ice cream only) $= 1/6 - 1/30 = 4/30$
        P(chips or ice cream but not both) $= 5/30 + 4/30 = 3/10$

3.4.3 $P(A) = 1/3$ $P(B) = 1/2$ $P(C) = 2/3$
        $P(A \cap B) = 1/6$ $P(A \cap C) = 2/9$ $P(B \cap C) = 1/3$ $P(A \cap B \cap C) = 2/9$

3.4.4 $P(M) = 1/2$ $P(N) = 1/2$ $P(Q) = 1/2$
        $P(M \cap N) = 1/4$ $P(M \cap Q) = 1/4$ $P(N \cap Q) = 1/4$ $P(M \cap N \cap Q) = 1/8$

3.4.5 $P(A) = 1/2$ $P(B) = 1/2$ $P(C) = 1/2$
        $P(A \cap B) = 1/4$ $P(A \cap C) = 1/4$ $P(B \cap C) = 1/4$ $P(A \cap B \cap C) = 0$

3.4.6 $P(S) = 1/2$ $P(T) = 1/2$ $P(U) = 1/2$
        $P(S \cap T) = 1/4$ $P(S \cap U) = 3/8$ $P(T \cap U) = 1/8$ $P(S \cap T \cap U) = 1/8$

3.5.5   (i) $(2/3)^{10} = 1024/59049$ (ii) $1 - (1024/59049) = 58025/59049$
        (iii) $10 \times 1/3 \times (2/3)^9 = 5120/59049$

3.5.6   (i) $(1/2)^3 = 1/8$ (ii) $(1/2)^3 = 1/8$
        (iii) $3 \times (1/2) \times (1/2)^2 = 3/8$ (iv) $3 \times (1/2) \times (1/2)^2 = 3/8$

3.5.7   (i) $(2/5)^4 = 16/625$ (ii) $(3/5)^4 = 81/625$
        (iv) $^4C_2 \times (2/5)^2 \times (3/5)^2 = 216/625$

3.5.8   (a) (i) $1/6$ (ii) $5/6 \times 1/6 = 5/36$
        (iii) $(5/6)^4 \times 1/6 = 625/7776$
        (iv) $\left(\dfrac{5}{6}\right)^{2n} < 1/1000 \Rightarrow n > 18.9 \Rightarrow n = 19$

3.5.9 (i) 1/4 (ii) $(3/4)^3 \times 1/4 = 27/256$
(iii) $(3/4)^6 \times 1/4$
(iv) $1/4 + (3/4)^3 \times 1/4 + (3/4)^6 \times 1/4$
(v) $1/4 + (3/4)^3 \times 1/4 + (3/4)^6 \times 1/4 + \ldots + (3/4)^{27} \times 1/4$
(vi) $1/4 + (3/4)^3 \times 1/4 + (3/4)^6 \times 1/4 + \ldots$

$$= \frac{1/4}{1 - (3/4)^3} = 16/37 \text{ (Infinite G.P.)}$$

Answers to self-assessment questions (Module 4)

4.1.3 Either $\dfrac{{}^6C_2 \times {}^4C_1}{{}^{10}C_3} = \dfrac{15 \times 4}{120} = \dfrac{1}{2}$

or $3 \times 6/10 \times 5/9 \times 4/8 = 360/720 = 1/2$

4.1.5 Let WE be the event that I wake up early,
WL the event that I wake up late,
AE the event that I arrive early,
and AL the event that I arrive late.

$P(AL) = P(WE \cap AL) + P(WL \cap AL)$
$= 12/150 + 20/150$
$= 32/150 = 16/75$

4.1.6

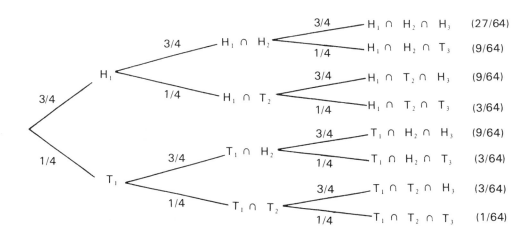

P(one head) = $P(H_1 \cap T_2 \cap T_3) + P(T_1 \cap H_2 \cap T_3) + P(T_1 \cap T_2 \cap H_3)$
= 3/64 + 3/64 + 3/64
= 9/64

4.1.7 Let K be the event that a person has previous knowledge of German and Pa that a person passes the examination.

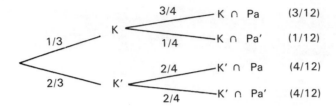

$P(Pa) = P(K \cap Pa) + P(K' \cap Pa) = 3/12 + 4/12 = 7/12$

4.1.8

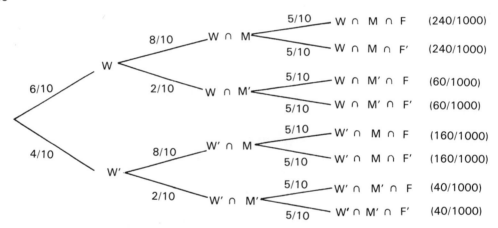

P(at least two A levels) = $P(W \cap M \cap F) + P(W \cap M \cap F') + P(W \cap M' \cap F) + P(W' \cap M \cap F)$
= 240/1000 + 240/1000 + 60/1000 + 160/1000
= 700/1000 = 0.7

4.1.9

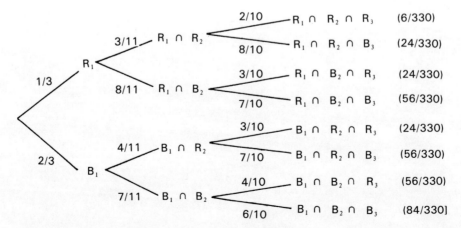

P(at least two blue marbles) = $P(R_1 \cap B_2 \cap B_3) + P(B_1 \cap R_2 \cap B_3) + P(B_1 \cap B_2 \cap R_3) + P(B_1 \cap B_2 \cap B_3)$
= 56/330 + 56/330 + 56/330 + 84/330 = 252/330 = 42/55

4.1.10

```
              3/7      A ∩ R    (6/21)
        A <
  2/3  /    4/7        A ∩ W    (8/21)
      <
  1/3  \    5/7        B ∩ R    (5/21)
        B <
              2/7      B ∩ W    (2/21)
```

(i) P(R) = P(A∩R) + P(B∩R)
    = 6/21 + 5/21 = 11/21

(ii)
$$P(A|R) = \frac{6/21}{6/21 + 5/21} = \frac{6}{6+5} = 6/11$$

4.1.11

```
              3/120    A ∩ B    (9/720)
        A <
  3/6  /    117/120    A ∩ B'   (351/720)
      /
      /       4/120    E ∩ B    (8/720)
  2/6 ——— E <
      \    116/120     E ∩ B'   (232/720)
      \
  1/6  \      6/120    D ∩ B    (6/720)
        D <
             114/120   D ∩ B'   (114/720)
```

(i) P(B) = P(A∩B) + P(E∩B) + P(D∩B) = 9/720 + 8/720 + 6/720 = 23/720

(ii)
$$P(E|B') = \frac{230/720}{351/720 + 232/720 + 114/720}$$

$$= \frac{232}{351 + 232 + 114} = 232/697$$

4.2.3

4/5 × 1/10 + 1/5 × 2/3 = 4/5 × 3/30 + 1/5 × 20/30
   = 32/150 = 16/75

4.2.4 2/3 × 1/2 + 1/3 × 3/4 = 1/3 + 1/4 = 7/12

4.2.5 (i) 2/3 x 3/7 + 1/3 x 5/7 = 6/21 + 5/21 = 11/21
       (ii) (6/21)/(11/21) = 6/11

4.2.6 (i) 1/2 x 1/40 + 1/3 x 1/30 + 1/6 x 1/20
            = 1/80 + 1/90 + 1/120
            = 9/720 + 8/720 + 6/720 = 23/720

       (ii) $\dfrac{1/3 \times 29/30}{697/720} = \dfrac{232/720}{697/720} = \dfrac{232}{697}$

4.2.7 3/6 x 1/5 + 2/6 x 1/6 + 1/6 x 1/8 = 3/6 x 24/120 + 2/6 x 20/120 + 1/6 x 15/120 = 127/720

4.2.8 1/4 x 1/10 + 1/3 x 1/5 + 5/12 x 1/20
       = 1/40 + 1/15 + 1/48
       = 6/240 + 16/240 + 5/240 = 27/240 = 9/80

4.2.9 (i) 3/10 x 2/5 + 7/20 x 1/2 + 1/4 x 7/10 + 1/10 x 3/10
            = 24/200 + 35/200 + 35/200 + 6/200
            = 100/200 = 1/2 = 50%

       (ii) $\dfrac{1/4 \times 7/10}{1/2} = \dfrac{7/40}{20/40} = \dfrac{7}{20}$

4.3.5 $\dfrac{1/3 \times 3/5}{1/3 \times 3/7 + 1/3 \times 4/9 + 1/3 \times 3/5} = \dfrac{3/5}{3/7 + 4/9 + 3/5}$

       $= \dfrac{189/315}{135/315 + 140/315 + 189/315} = \dfrac{189}{464}$

4.3.6 $\dfrac{1/10 \times 3/100}{3/10 \times 2/100 + 6/10 \times 1/100 + 1/10 \times 3/100} = \dfrac{3}{6 + 6 + 3} = \dfrac{1}{5}$

4.3.7 $\dfrac{1/10000 \times 95/100}{1/10000 \times 95/100 + 9999/10000 \times 1/100} = \dfrac{95}{95 + 9999} = \dfrac{95}{10094}$

4.3.8 $\dfrac{2/7 \times 3/4}{2/7 \times 3/4 + 5/7 \times 1/2} = \dfrac{6}{6 + 10} = \dfrac{3}{8}$

4.3.9 $\dfrac{1/3 \times 5/10}{1/3 \times 5/10 + 2/3 \times 6/10} = \dfrac{5}{5 + 12} = \dfrac{5}{17}$

4.3.10 $\dfrac{9/10 \times 7/10}{9/10 \times 7/10 + 1/10 \times 4/10} = \dfrac{63}{63 + 4} = \dfrac{63}{67}$

4.3.11 $\dfrac{2/3 \times 1/5}{2/3 \times 1/5 + 1/3 \times 1} = \dfrac{2}{2 + 5} = \dfrac{2}{7}$

4.3.12 $\quad \dfrac{4/10 \times 6/100}{2/10 \times 8/100 + 4/10 \times 6/100 + 3/10 \times 4/100 + 1/10 \times 3/100}$

$= \dfrac{24}{16 + 24 + 12 + 3} = \dfrac{24}{55}$

SOLUTIONS

Module 1
TYPE 1	1. (c)	2. (b)	3. (b)
TYPE 2	1. F	2. F	3. F
TYPE 3	1. (a) (b)	2. (a)	3. (a) (b) (c)
TYPE 4	1. D	2. B	3. C
TYPE 5	1. (b)	2. A	3. I

Module 2
TYPE 1	1. (d)	2. (c)	3. (a)
TYPE 2	1. T	2. F	3. T
TYPE 3	1. (a) (c)	2. (b)	3. (a) (b) (c)
TYPE 4	1. E	2. D	3. C
TYPE 5	1. I	2. (b)	3. A

Module 3
TYPE 1	1. (d)	2. (b)	3. (d)
TYPE 2	1. F	2. T	3. T
TYPE 3	1. (a) (b)	2. (a)	3. (a) (b)
TYPE 4	1. D	2. A	3. C
TYPE 5	1. A	2. I	3. (a) (c)

Module 4
TYPE 1	1. (a)	2. (c)	3. (c)
TYPE 2	1. F	2. F	3. T
TYPE 3	1. (a) (c)	2. (a) (b) (c)	3. (a) (b)
TYPE 4	1. B	2. A	3. D
TYPE 5	1. I	2. A	3. (c)